Praise for Oliver Sacks's

On the Move

"A beautiful vision, one that embraces an infinite spectrum of wonder. . . . *On the Move* illustrates what an exceptional human being [Sacks] is. . . . He is fascinated by seemingly everything, and, damn, the man can write."　　—*Salon*

"Marvelous. . . . He studies himself as he has studied others: compassionately, unblinkingly, intelligently, acceptingly and honestly."　　　　　　　　　　—*The Wall Street Journal*

"Sacks's ability to enact and celebrate intuition in medicine and precision in art is singular."
　　　　　　　　　　　—*The New York Times Book Review*

"[Sacks is] a wonderful storyteller. . . . It's his keen attentiveness as a listener and observer, and his insatiable curiosity, that makes his work so powerful."
　　　　　　　　　　　　　　—*San Francisco Chronicle*

"Remarkably candid and deeply affecting. . . . Sacks's empathy and intellectual curiosity, his delight in, as he calls it, 'joining particulars with generalities' and, especially, 'narratives with neuroscience'—have never been more evident than in his beautifully conceived new book."
　　　　　　　　　　　　　　　　—*The Boston Globe*

"This remarkable man lifts us all. . . . [*On the Move*] is not only a record of his life-affirming characterological extravagance but also a meditation on what it is to be human in an age of medical arrogance and the numbing clout of technology."
—*Los Angeles Review of Books*

"A compelling read. . . . Offers a glimpse into one of the greatest minds of our time."
—*Men's Journal*

"What a self this book reveals! A man animated by boundless curiosity, wide-ranging intelligence, gratitude for flawed humanity, perseverance despite setbacks. . . . We're lucky to have all the books, including *On the Move*. It's intensely, beautifully, incandescently alive."
—*Newsday*

"An ebullient telling of a remarkable life."
—*Paste*

"Intriguing. . . . When describing his patients and their problems, he is attentive and precise, straightforward and sympathetic, and he brings these worthy qualities to his descriptions of his younger self."
—*The Washington Post*

"An unforgettably passionate, joyous journey."
—*The Daily Beast*

"[A] beautifully constructed and moving memoir. . . . His life and work are a gift."
—*The Times Literary Supplement* (London)

"Moving. . . . Written with exceptional grace and clarity."
—*Richmond Times-Dispatch*

Oliver Sacks

On the Move

Oliver Sacks was a physician, writer, and professor of neurology. Born in London in 1933, he moved to New York City in 1965, where he launched his medical career and began writing case studies of his patients. Called the "poet laureate of medicine" by *The New York Times*, Sacks is the author of more than a dozen books, including *The Man Who Mistook His Wife for a Hat*, *Musicophilia*, and *Awakenings*, which inspired an Oscar-nominated film and a play by Harold Pinter. He was the recipient of many awards and honorary degrees and was made a Commander of the British Empire in 2008 for services to medicine. He died in 2015.

www.oliversacks.com

On the Move

On the Move

A Life

Oliver Sacks

Vintage Books
A Division of Penguin Random House LLC
New York

FIRST VINTAGE BOOKS EDITION, FEBRUARY 2016

The Library of Congress has cataloged the Knopf edition as follows:
Sacks, Oliver W.
On the move : a life / by Oliver Sacks.
pages cm
1. Sacks, Oliver W.   2. Neurologists—England—biography.
3. Neurologists—United States—biography.   I. Title.
RC339.52.S23A3 2015   616.80092—dc23   [B]   2015001870

Vintage Books Trade Paperback ISBN: 978-0-8041-7093-2
eBook ISBN: 978-0-385-35255-0

Book design by Iris Weinstein

www.vintagebooks.com

Printed in the United States of America
10 9 8 7 6 5 4 3 2 1

for Billy

"Life must be lived forwards but can only be understood backwards."

—*Kierkegaard*

Contents

On the Move

On the Move

When I was at boarding school, sent away during the war as a little boy, I had a sense of imprisonment and powerlessness, and I longed for movement and power, ease of movement and superhuman powers. I enjoyed these, briefly, in dreams of flying and, in a different way, when I went horse riding in the village near school. I loved the power and suppleness of my horse, and I can still evoke its easy and joyous movement, its warmth and sweet, hayey smell.

Most of all, I loved motorbikes. My father had had one before the war, a Scott Flying Squirrel with a big water-cooled engine and an exhaust like a scream, and I wanted a powerful bike, too. Images of bikes and planes and horses merged for me, as did images of bikers and cowboys and pilots, whom I imagined to be in precarious but jubilant control of their powerful mounts. My boyish imagination was fed by Westerns and films of heroic air combat, seeing pilots risking their lives in Hurricanes and Spitfires but lent protection by their thick flying jackets, as motorcyclists were by their leather jackets and helmets.

When I returned to London as a ten-year-old in 1943, I enjoyed sitting in the window seat of our front room, watching and trying to identify motorbikes as they sped by (after the war, when petrol was easier to get, they became much commoner). I could identify a dozen or more marques—AJS, Triumph, BSA, Norton, Matchless, Vincent, Velocette, Ariel, and Sunbeam, as well as rare foreign bikes like BMWs and Indians.

As a teenager, I would go regularly to Crystal Palace with a like-minded cousin to see the motorbike racing there. I often hitchhiked to Snowdonia to climb or to the Lake District to swim and sometimes got a lift on a motorbike. Riding pillion thrilled me and stimulated daydreams of the sleek, powerful bike I would get one day.

My first motorbike, when I was eighteen, was a secondhand BSA Bantam with a little two-stroke engine and, as it turned out, faulty brakes. I took it to Regent's Park on its maiden ride, which turned out to be fortunate, possibly lifesaving, because the throttle jammed when I was going flat out and the brakes were not strong enough to stop the bike or even slow it more than a little. Regent's Park is encircled by a road, and I found myself going round and round it, perched on a motorbike I had no way of stopping. I hooted or yelled to warn pedestrians out of my way, but after I had made two or three circuits, everyone gave me a free path and shouted encouragement as I passed by again and again. I knew the bike would have to stop eventually, when it ran out of gas, and finally, after dozens of involuntary circuits of the park, the engine sputtered and died.

My mother had been very much against my getting a bike in the first place. That I expected, but I was surprised by my father's opposition, since he had ridden a bike himself. They had tried to dissuade me from getting a bike by buying me a little car, a 1934 Standard that could barely do forty miles per hour. I had grown to hate the little car, and one day, impulsively, I sold it and used the proceeds to buy the Bantam. Now I had to explain to my parents that a feeble little car or bike was dangerous because one lacked the power to pull out of trouble and that I would be much safer with a larger, more powerful bike. They acceded to this reluctantly and funded me for a Norton.

On my first Norton, a 250 cc machine, I had a couple of near accidents. The first came when I approached a red traffic light too fast and, realizing that I could not safely brake or turn, drove straight on and somehow—miraculously—passed between two lines of cars going in opposite directions. Reaction came a minute later: I rode another block, parked the bike in a side road—and fainted.

The second accident occurred at night in heavy rain on a winding country road. A car coming in the opposite direction did not dim its headlights, and I was blinded. I thought there would be a head-on collision, but at the last moment I stepped off the bike (an expression of ridiculous mildness for a potentially lifesaving but potentially fatal maneuver). I let the bike go in one direction (it missed the car but was totaled) and myself in another. Fortunately, I was wearing a helmet, boots, and gloves, as well as full leathers, and though I slid twenty yards or so on the rain-slicked road, I was so well protected by my clothing that I did not get a scratch.

My parents were shocked, but very glad I was in one piece, and raised strangely little objection to my getting another, more powerful bike—a 600 cc Norton Dominator. At this point, I had finished at Oxford, and I was about to go to Birmingham, where I had a job as a house surgeon for the first six months of 1960, and I was careful to say that with the newly opened M1 motorway between Birmingham and London and a fast bike, I would be able to come home every weekend. The motorway in those days had no speed limit, so I could be back in a little over an hour.

I met up with a motorcycle group in Birmingham and tasted the pleasure of being part of a group, sharing an enthusiasm; up to this point, I had always been a solitary rider. The countryside around Birmingham was quite unspoiled, and a special pleasure was riding to Stratford-on-Avon to see whatever Shakespeare play was on.

In June of 1960, I went to the TT, the great Tourist Trophy motorcycle race held annually on the Isle of Man. I managed to procure an Emergency Medical Service armband, which enabled me to visit the pits and see some of the riders. I kept careful notes and had plans to write a motorbike-racing novel set on the Isle of Man—I did a great deal of research for this— but it never got off the ground.[1]

•

The North Circular Road around London also had no speed limits in the 1950s—very inviting for those who loved speed— and there was a famous café, the Ace, which was basically a hangout for motorcyclists with fast machines. "Doing the ton"—a hundred miles per hour—was a minimum criterion for being one of the inner group, the Ton-Up Boys.

A number of bikes, even then, could do the ton, especially if they were tuned up a little, relieved of surplus weight (including exhausts), and given high-octane fuel. More challenging was the "burn-up," a race around secondary roads, and you risked a challenge as soon as you entered the café. "Playing chicken," however, was discountenanced; the North Circular, even then, carried heavy traffic at times.

---

1. In a notebook I kept at the time, I indicated my intention to write five novels (including the motorcycle one), as well as a memoir on my chemical boyhood. I never wrote the novels, but forty-five years later I wrote the memoir, *Uncle Tungsten*.

I never played chicken, but I enjoyed a little road racing; my 600 cc "Dommie" had a slightly souped-up engine but could not match the 1000 cc Vincents favored by the inner circle at the Ace. I once tried a Vincent, but it seemed horribly unstable to me, especially at low speeds, very different from my Norton, which had a "feather bed" frame and was wonderfully stable, whatever one's speed. (I wondered if one could fit a Vincent engine in a Norton frame, and I was to find, years later, that such "Norvins" had been made.) When speed limits were introduced, there was no more doing the ton; the fun was over, and the Ace ceased to be the place it once was.

When I was twelve, a perceptive schoolmaster wrote in his report, "Sacks will go far, if he does not go too far," and this was often the case. As a boy, I often went too far in my chemical experiments, filling the house with noxious gases; luckily, I never burned the place down.

I liked to ski, and when I was sixteen, I went to Austria with a school group for some downhill skiing. The following year I traveled alone to do cross-country skiing in Telemark. The skiing went well, and before taking the ferry back to England, I bought two liters of aquavit in the duty-free shop and then went through Norwegian border control. As far as the Norwegian customs officers were concerned, I could have any number of bottles with me, but (they informed me) I could bring only one bottle *into* England; U.K. customs would confiscate the other. I got on board, clutching my two bottles, and made for the upper deck. It was a brilliantly clear, very cold day, but having all my warm ski clothes with me, I did not see this as a problem; everyone else stayed inside, and I had the entire upper deck to myself.

I had my book to read—I was reading *Ulysses,* very slowly—and my aquavit to sip: nothing like alcohol to warm one inside. Lulled by the gentle, hypnotic motion of the ship, taking a little aquavit from time to time, I sat on the upper deck, absorbed in my book. I was surprised to find, at one point, that I had drunk, in tiny increments, almost half the bottle. I noticed no effect, so I continued reading and sipping from the bottle, increasingly upended now it was half-empty. I was rather startled when I realized we were docking; I had been so absorbed by *Ulysses* that I failed to note the passage of time. The bottle was now empty. I still felt no effects; the stuff must be much weaker than they make out, I thought, even though it said "100 proof" on the label. I noticed nothing amiss, until I stood up and promptly fell flat on my face. I was extremely surprised by this—had the ship suddenly lurched? So I got up and immediately fell down once again.

Only now it began to dawn on me that I was drunk—very, very drunk—though the drink had apparently gone straight to my cerebellum, leaving the rest of my head alone. Coming up to make sure everyone was off the boat, a crewman found me endeavoring to walk, using my ski poles for support. He called an assistant, and the two of them, one on each side, escorted me off the boat. Though lurching badly and attracting (mostly amused) attention, I felt I had beaten the system, leaving Norway with two bottles but arriving with one. I had cheated the U.K. customs of a bottle which, I imagined, they would dearly have liked for themselves.

Nineteen fifty-one was an eventful, and in some ways painful, year. My Auntie Birdie, who had been a constant presence in my life, died in March; she had lived with us for my entire lifetime and was unconditionally loving to us all. (Birdie was a tiny woman and of modest intelligence, the only one so handicapped among my mother's siblings. It was never quite clear to me what had happened to her in early life; there was talk of a head injury in infancy but also of a congenital thyroid deficiency. None of this mattered to us; she was simply Auntie Birdie, an essential part of the family.) I was greatly affected by Birdie's death and perhaps only then realized how deeply she was woven into my life, all our lives. When, a few months before, I got a scholarship to Oxford, it was Birdie who gave me the telegram and hugged and congratulated me—shedding some tears, too, because she knew this meant that I, the youngest of her nephews, would be leaving home.

I was due to go to Oxford in late summer. I had just turned eighteen, and my father thought this was the time for a serious man-to-man, father-to-son talk with me. We talked about allowances and money—not a big issue, for I was fairly frugal in my habits and my only extravagance was books. And then my father got on to what was really worrying him.

"You don't seem to have many girlfriends," he said. "Don't you like girls?"

"They're all right," I answered, wishing the conversation would stop.

"Perhaps you prefer boys?" he persisted.

"Yes, I do—but it's just a feeling—I have never 'done' anything," and then I added, fearfully, "Don't tell Ma—she won't be able to take it."

But my father did tell her, and the next morning she came down with a face of thunder, a face I had never seen before. "You are an abomination," she said. "I wish you had never been born." Then she left and did not speak to me for several days. When she did speak, there was no reference to what she had said (nor did she ever refer to the matter again), but something had come between us. My mother, so open and supportive in most ways, was harsh and inflexible in this area. A Bible reader like my father, she loved the Psalms and the Song of Solomon but was haunted by the terrible verses in Leviticus: "Thou shall not lie with mankind, as with womankind: it is abomination."

My parents, as physicians, had many medical books, including several on "sexual pathology," and I had dipped into Krafft-Ebing, Magnus Hirschfeld, and Havelock Ellis by the age of twelve. But I found it difficult to feel that I had a "condition," that my identity could be reduced to a name or a diagnosis. My friends at school knew that I was "different," if only because I excused myself from parties which would end in petting and necking.

Buried in chemistry and then in biology, I was not too aware of what was going on all around me—or inside me—and I had no crushes on anyone at school (although I was turned on by a full-size reproduction, at the head of the stairway, of the famous statue of a beautifully muscled, naked Laocoön trying to save his sons from the serpents). I knew that the very idea of homosexuality aroused horror in some people; I suspected that this might be the case with my mother, which is why I said to my father, "Don't tell Ma—she won't be able to take it." I should not, perhaps, have told my father; in general, I regarded my sexuality as nobody's business but my own, not a secret, but not to be talked about. My closest friends, Eric and Jonathan, were aware of it, but we almost never discussed the subject. Jonathan said that he regarded me as "asexual."

We are all creatures of our upbringings, our cultures, our times. And I have needed to remind myself, repeatedly, that my mother was born in the 1890s and had an Orthodox upbringing and that in England in the 1950s homosexual behavior was treated not only as a perversion but as a criminal offense. I have to remember, too, that sex is one of those areas—like religion and politics—where otherwise decent and rational people may have intense, irrational feelings. My mother did not mean to be cruel, to wish me dead. She was suddenly overwhelmed, I now realize, and she probably regretted her words or perhaps partitioned them off in a closeted part of her mind.

But her words haunted me for much of my life and played a major part in inhibiting and injecting with guilt what should have been a free and joyous expression of sexuality.

My brother David and his wife, Lili, learning of my lack of sexual experience, felt it could be attributed to shyness and that a good woman, even a good fuck, could set me to rights. Around Christmas of 1951, after my first term at Oxford, they took me to Paris with the intention not only of seeing the sights—the Louvre, Notre Dame, the Eiffel Tower—but of taking me to a kindly whore who would put me through my paces, skillfully and patiently teaching me what sex was like.

A prostitute of suitable age and character was selected—David and Lili interviewed her first, explaining the situation—and I then went into her room. I was so frightened that my penis became limp with fear and my testicles tried to retreat into my abdominal cavity.

The prostitute, who resembled one of my aunts, saw the situation at a glance. She spoke good English (this had been one of the criteria for her selection), and she said, "Don't worry— we'll have a nice cup of tea instead." She pulled out tea things and petits fours, put on a kettle, and asked what sort of tea I liked. "Lapsang," I said. "I love the smoky smell." By this time, I had recovered my voice and my confidence and chatted easily with her as we enjoyed our smoky tea.

I stayed for half an hour, then left; my brother and his wife were waiting, expectantly, outside. "How was it, Oliver?" David demanded. "Terrific," I said, wiping crumbs off my beard.

B y the time I was fourteen, it was "understood" that I was going to be a doctor. My mother and father were both physicians, and so were my two eldest brothers.

I was not sure, however, that I wanted to be a doctor. I could no longer nourish ambitions to be a chemist; chemistry itself had advanced beyond the eighteenth- and nineteenth-century inorganic chemistry I loved so much. But at fourteen or fifteen, inspired by my school biology teacher and by Steinbeck's *Cannery Row*, I thought I would like to become a marine biologist.

When I got my scholarship to Oxford, I faced a choice: Should I stick to zoology or become a pre-med student and do anatomy, biochemistry, and physiology? It was especially the physiology of the senses that fascinated me—how did we see color, depth, movement? How did we *recognize* anything? How did we make sense of the world, visually? I had developed these interests from an early age through having visual migraines, for besides the brilliant zigzags which heralded an attack, I might, during a migraine aura, lose the sense of color or depth or movement or even the ability to recognize anything. My vision could be unmade, deconstructed, frighteningly but fascinatingly, in front of me, and then be remade, reconstructed, all in the space of a few minutes.

My little home chemistry lab had doubled as a photography darkroom, and I was especially drawn to color and stereophotography; these too made me wonder how the brain constructed color and depth. I had enjoyed marine biology much as I had enjoyed chemistry, but now I wanted to understand how the human brain worked.

.

I never had much intellectual self-confidence, even though I was regarded as bright. Like my two closest school friends, Jonathan Miller and Eric Korn, I was obsessed with both science and literature. I was in awe of Jonathan's and Eric's intelligence and couldn't think why they hung around with me, but we all got scholarships to university. I then ran into some difficulties.

At Oxford, one had to take an exam called "prelims" for entry; it was considered a mere formality with me, because I already had an open scholarship. But I failed prelims; I took them a second time, and I failed again. I took the test a third time and failed yet again, and at this point Mr. Jones, the Provost, pulled me aside and said, "You did splendid scholarship papers, Sacks. Why are you failing this silly exam again and again?" I said I didn't know, and he said, "Well, this is your last chance." So I took the test a fourth time and finally passed.

At St. Paul's School, with Eric and Jonathan, I could enjoy an easy mix of arts and sciences. I was president of our literary society and secretary of the Field Club at the same time. Such a mix was more difficult at Oxford, for the anatomy department, the science laboratories, and the Radcliffe Science Library were all clustered together in South Parks Road, at a distance from the university lecture halls and colleges. There was both a physical and a social separation between those of us doing science or pre-med degrees and the rest of the university.

I felt this sharply in my first term at Oxford. We had to write essays and present these to our tutors, and this entailed many hours in the Radcliffe Science Library, reading research and review papers, culling what seemed most important, and presenting it in an interesting and individual way. Spending a great deal of time reading neurophysiology was enjoyable, even thrilling—vast new areas seemed to be opening out—but I became more and more conscious of what was now missing from my life. I was doing almost no general reading other than Maynard Keynes's *Essays in Biography*, and I wanted to write my own "Essays in Biography," though with a clinical twist—essays presenting individuals with unusual weaknesses or strengths and showing the influence of these special features on their lives; they would, in short, be clinical biographies or case histories of a sort.

My first (and, in the event, my only) subject here was Theodore Hook, whose name I had come across while reading a biography of Sydney Smith, the great early Victorian wit. Hook too was a great wit and conversationalist, a decade or two earlier than Sydney Smith; he also had, to an unrivaled degree, powers of musical invention. It was said that he had composed more than five hundred operas, sitting at a piano, improvising, and singing all of the parts. These were flowers of the moment—astonishing, beautiful, and ephemeral; they were improvised on the spot, never repeated, never written down, and soon forgotten. I was enthralled by descriptions of Hook's improvisational genius; what sort of brain could allow for this?

I started reading what I could about Hook, as well as some of the books he had written; they seemed oddly dull and labored, in contrast to descriptions of his lightning-quick, wildly inventive improvisations. I thought about Hook a good deal, and towards the end of the Michaelmas term I wrote an essay on him, an essay which ran to six closely typed foolscap pages, four or five thousand words in all.

I recently found this essay in a box, along with other early writings. Reading it, I am struck by its fluency, its erudition, its pomposity, and its pretentiousness. It does not seem like my writing. Could I have cribbed the entire thing or stitched it together from half a dozen sources, or was it in fact my own writing, couched in a learned, professorial style which I had adopted to counter the fact that I was a callow eighteen-year-old?

Hook was a diversion; most of my essays were on physiological subjects and were to be read weekly to my tutor. When I took on the subject of hearing, I got so excited by this, did so much reading and thinking, that I did not actually have time to write my essay. But on the day of my presentation, I brought in a pad of paper and pretended to read from it, turning over the pages as I extemporized on the subject. At one point, Carter (Dr. C. W. Carter, my tutor at Queen's) stopped me.

"I didn't quite follow that," he said. "Could you read it again?" A little nervously, I tried to repeat the last couple of sentences. Carter looked puzzled. "Let me see it," he said. I handed him the blank pad. "Remarkable, Sacks," he said. "Very remarkable. But in future, I want you to *write* your essays."

•

As a student at Oxford, I had access not only to the Radcliffe Science Library but to the Bodleian, a wonderful general library that could trace its origins back to 1602. It was in the Bodleian that I stumbled upon Hook's now obscure and forgotten works. No other library—apart from the British Museum Library— could have provided the materials I needed, and the tranquil atmosphere of the Bodleian was a perfect one in which to write.

But the library I most loved at Oxford was our own library at the Queen's College. The magnificent library building, we were told, had been designed by Christopher Wren, and beneath this, in an underground maze of heating pipes and shelves, were the vast subterranean holdings of the library.

To hold ancient books, incunabula, in my own hands was a new experience for me; I particularly adored Conrad Gesner's *Historiae animalium* (1551), richly illustrated (it had Albrecht Dürer's famous drawing of a rhinoceros), and Louis Agassiz's four-volume work on fossil fish. It was in the stacks that I saw all of Darwin's works in their original editions, and there, too, that I fell in love with all the works of Sir Thomas Browne— his *Religio Medici*, his *Hydriotaphia*, and *The Garden of Cyrus* (*The Quincunciall Lozenge*). How absurd some of these were, but how magnificent the language! And if Browne's classical magniloquence became too much at times, one could switch to the lapidary cut and thrust of Swift, all of whose works, of course, were there in their original editions. While I had grown up on the nineteenth-century works that my parents favored, it was the catacombs of the Queen's library that introduced me to seventeenth- and eighteenth-century literature—Johnson, Hume, Gibbon, and Pope. All of these books were freely available, not in some special, locked-away, rare books enclave, but just sitting on the shelves, as they had done, I imagined, since their original publication. It was in the vaults of the Queen's College that I really gained a sense of history and of my own language.

•

My mother, a surgeon and anatomist, while accepting that I was too clumsy to follow in her footsteps as a surgeon, expected me at least to excel in anatomy at Oxford. We dissected bodies and attended lectures and, a couple of years later, had to sit for a final anatomy exam. When the results were posted, I saw that I was ranked one from bottom in the class. I dreaded my mother's reaction and decided that, in the circumstances, a few drinks were called for. I made my way to a favorite pub, the White Horse in Broad Street, where I drank four or five pints of hard cider—stronger than most beer and cheaper, too.

Rolling out of the White Horse, liquored up, I was seized by a mad and impudent idea. I would try to compensate for my abysmal performance in the anatomy finals by having a go at a very prestigious university prize—the Theodore Williams Scholarship in Human Anatomy. The exam had already started, but I lurched in, drunkenly bold, sat down at a vacant desk, and looked at the exam paper.

There were seven questions to be answered; I pounced on one ("Does structural differentiation imply functional differentiation?") and wrote nonstop for two hours on the subject, bringing in whatever zoological and botanical knowledge I could muster to flesh out the discussion. Then I left, an hour before the exam ended, ignoring the other six questions.

The results were in *The Times* that weekend; I, Oliver Wolf Sacks, had won the prize. Everyone was dumbfounded—how could someone who had come one but last in the anatomy finals walk off with the Theodore Williams prize? I was not entirely surprised, for it was a sort of repetition, in reverse, of what had happened when I took the Oxford prelims. I am very bad at factual exams, yes-or-no questions, but can spread my wings with essays.

Fifty pounds came with the Theodore Williams prize—£50! I had never had so much money at once. This time I went not to the White Horse but to Blackwell's bookshop (next door to the pub) and bought, for £44, the twelve volumes of the *Oxford English Dictionary*, for me the most coveted and desirable book in the world. I was to read the entire dictionary through when I went on to medical school, and I still like to take a volume off the shelf, now and then, for bedtime reading.

wow

•

My closest friend at Oxford was a Rhodes scholar, a young mathematical logician called Kalman Cohen. I had never met a logician before, and I was fascinated by Kalman's power of intellectual focus. He seemed able to fix his mind on a problem nonstop for weeks on end, and he had a passion for thinking; the very act of thinking seemed to excite him, irrespective of the thoughts he arrived at.

Though we were so different, we got on very well. He was attracted by my sometimes wildly associative mind, as I was by his highly focused mind. He introduced me to Hilbert and Brouwer, the giants of mathematical logic, and I introduced him to Darwin and the great nineteenth-century naturalists.

We think of science as discovery, art as invention, but is there a "third world" of mathematics, which is somehow, mysteriously, both? Do numbers—primes, for example—exist in some eternal Platonic realm? Or were they invented, as Aristotle thought? What is one to make of irrational numbers, like π? Or imaginary numbers, like the square root of -2? Such questions exercised me, fruitlessly, from time to time, but they were almost a life-and-death matter for Kalman. His hope was to somehow reconcile Brouwer's Platonic intuitionism with Hilbert's Aristotelian formalism, their so different yet complementary views of mathematical reality.

When I spoke of Kal to my parents, they immediately thought of how far he was from home and invited him to spend a relaxed weekend, with home cooking, in our house in London. My parents enjoyed meeting him, but my mother was indignant the next morning when she found one of Kal's bedsheets covered with inky writing. When I explained that he was a genius and that he had used the sheet to work out a new theory in mathematical logic (here I exaggerated a little), her indignation changed to awe, and she insisted on keeping the sheet, unwashed, unerased, in case, on a future visit, Kalman might want to consult it again. She also showed it proudly to Selig Brodetsky, a former Senior Wrangler at Cambridge (and an ardent Zionist), the only mathematician she knew.

Kalman had been at Reed College in Oregon—this, he told me, was known for its brilliant students—and he had been its highest-ranking graduate in many years. He said this simply, unaffectedly, as one might speak of the weather. It was simply a matter of fact. He seemed to think I was bright, too, despite the manifest disorder and illogic of my mind. He felt that bright people should marry each other and have bright children, and with this in mind he arranged for me to meet another Rhodes scholar from America, a Miss Isaac. Rael Jean was quiet, self-effacing, but (as Kal had said) diamond sharp, and we talked high abstractions all through our dinner together. We parted amicably but never saw each other again, nor did Kalman attempt to find me a mate again.

In the summer of 1952, our first long vac, Kalman and I hitchhiked through France to Germany, sleeping at youth hostels on the way. Somewhere we picked up head lice and had to have our heads shaved. A rather elegant friend at Queen's College, Gerhart Sinzheimer, had invited us to stop by; he was summering with his parents in their house on the Titisee in the Black Forest. When Kalman and I arrived, filthy and bald-headed with a story of catching lice, they ordered us both to have a bath, and they had our clothes fumigated. After a short, awkward stay with the elegant Sinzheimers, we made our way to Vienna (then very much, we thought, the Vienna of *The Third Man*), and there we sampled every liqueur known to man.

·

Although I was not taking my degree in psychology, I sometimes went to lectures in the psychology department. It was there that I saw J. J. Gibson, a bold theorist and experimenter in visual psychology who had come to Oxford on a sabbatical from Cornell. Gibson had recently published his first book, *The Perception of the Visual World*, and was happy to let us experiment with special glasses that inverted (in one eye or both) what one normally saw. Nothing was more bizarre than seeing the world upside down, and yet, within days, the brain would adapt to this and reorient the visual world (only to have it appear upside down again when one took off the spectacles).

Visual illusions, too, fascinated me; they showed how intellectual understanding, insight, and even common sense were powerless against the force of perceptual distortions. Gibson's inverting glasses showed the power of the mind to rectify optical distortions, where visual illusions showed its inability to correct perceptual ones.

R ichard Selig. It has been sixty years, but I can still see Rich-ard's face, his bearing—he bore himself like a lion—as I first saw him outside Magdalen College in Oxford in 1953. We got talking; I suspect that it was he who started a conversation, for I was always too shy to initiate any contact and his great beauty made me even shyer. I learned in that first conversa-tion that he was a Rhodes scholar and a poet and had worked at a variety of odd jobs all over the States. His knowledge of the world was far greater than mine, even given the disparity of age (he was twenty-four; I was twenty), far greater than that of most undergraduates who had gone straight from school to university with no experience of real life in between. He found something interesting in me, and we soon became friends—and more, for I fell in love with him. It was the first time in my life I had fallen in love.

I fell in love with his face, his body, his mind, his poetry, everything about him. He would often bring me just-written poems, and I would give him some of my physiology essays in return. I was not, I think, the only one to fall in love with him; there were others, both men and women—his great beauty, his great gifts, his vitality and love of life, ensured this. He talked freely of himself—about his apprenticeship with the poet Theodore Roethke, his friendships with many painters, and the year he himself had spent as a painter before realizing that whatever his talents were, his real passion was for poetry. He would often carry images, words, lines of poetry, in his head, working on them consciously and unconsciously for months on end until they were born as finished poems or abandoned. He had had poems published in *Encounter*, *The Times Liter-ary Supplement*, *Isis*, and *Granta* and had a great supporter in

Stephen Spender. I thought he was a genius, or a genius in the making.

We would go on long walks together, talking about poetry and science. Richard loved to hear me wax enthusiastic about chemistry and biology, and I lost my shyness when I did so. While I knew that I was in love with Richard, I was very apprehensive of admitting this; my mother's words about "abomination" had made me feel that I must not say what I was. But, mysteriously, wonderfully, being *in* love, and in love with a being like Richard, was a source of joy and pride to me, and one day, with my heart in my mouth, I told Richard that I was in love with him, not knowing how he would react. He hugged me, gripped my shoulders, and said, "I know. I am not that way, but I appreciate your love and love you too, in my own way." I did not feel rebuffed or brokenhearted. He had said what he had to say in the most sensitive way, and our friendship continued, made easier now by my relinquishing certain painful and hopeless longings.

I thought we might be lifelong friends, as perhaps he did, too. But one day he came to my digs, looking disturbed. He had noticed a swelling in one of his groins; he had paid no attention to it at first, thinking it would disappear, but it had grown larger and was getting uncomfortable. I was a pre-med student, he said, could I look at it? He pulled his trousers and underpants down, and there it was, the size of an egg, in his left groin. It was fixed and hard to the touch. Cancer was my immediate thought. I said to Richard, "You must see a doctor—it may need a biopsy—don't delay."

The gland was biopsied and diagnosed as lymphosarcoma; Richard was told that he could expect no more than two years of life. After telling me this, he never spoke to me again; I was the first to recognize the deadly import of his tumor, and perhaps he saw me now as a sort of messenger or symbol of death.

But he was determined to live as fully as he could in the time remaining to him; he married the Irish harpist and singer Mary O'Hara, went with her to New York, and died fifteen months later. He wrote much of his finest poetry in these last months.

One takes one's finals at Oxford after three years. I stayed on to do research, and for the first time at Oxford I found myself rather isolated, for almost all my contemporaries had left.

I had been offered a research position in the anatomy department after being awarded the Theodore Williams Scholarship but declined the offer, despite my admiration for the professor of anatomy, the very eminent and eminently approachable Wilfrid Le Gros Clark.

Le Gros Clark was a wonderful teacher who portrayed all of human anatomy from an evolutionary perspective and was known, at the time, for his role in exposing the Piltdown hoax. But I declined his offer because I had been seduced by a series of vivid lectures on the history of medicine given by the university reader in human nutrition, H. M. Sinclair.

I had always loved history and even in my boyhood chemical days wanted to know about the lives and personalities of chemists, the controversies and conflicts that sometimes accompanied new discoveries or theories. I wanted to see how chemistry unfolded as a human enterprise. And now, in Sinclair's lectures, it was the history of physiology, the ideas and personalities of physiologists, which came to life.

Friends, and even my own tutor at Queen's, tried to warn

me, to dissuade me from what they felt would be a mistake. But though I had heard rumors about Sinclair—nothing too specific, merely comments on his being a "peculiar" and somewhat isolated figure in the university; rumors, too, that the university was going to close down his lab—I was not to be dissuaded.

I realized my mistake as soon as I started at the LHN, the Laboratory of Human Nutrition.

Sinclair's knowledge, at least his historical knowledge, was encyclopedic, and he guided me to work on something I had only vaguely heard about. The jake paralysis, so-called, had caused crippling neurological damage during Prohibition, when drinkers, denied legal forms of alcohol, turned to a very strong alcoholic extract of Jamaica ginger, or "jake," which was freely available then as a "nerve tonic." When its potential for abuse became apparent, the government had it doctored with a very unpleasant-tasting compound, triorthocresyl phosphate, or TOCP. But this hardly deterred drinkers, and it soon became apparent that TOCP was in fact a grave, albeit slow-acting, nerve poison. By the time this was realized, more than fifty thousand Americans had suffered extensive and often irreversible nerve damage. Those affected showed a distinctive paralysis of the arms and legs and developed a peculiar, easily recognized gait, the "jake walk."

Exactly how TOCP caused nerve damage was still unknown, though there had been some suggestion that it especially affected the myelin sheaths of the nerves, and, Sinclair said, there were no known antidotes. He challenged me to develop an animal model of the disease. Here, with my love of invertebrates, I thought immediately of earthworms: they had giant myelinated nerve fibers, which mediated the worms' ability to curl up suddenly when they were hurt or threatened. These nerve fibers were relatively easy to study, and there would never be any problem getting as many worms as I wanted. I could supplement the earthworms, I thought, with chickens and frogs.

Once we had discussed my project, Sinclair secreted himself in his book-lined office and became virtually inaccessible—not only to me, but to everyone in the Laboratory of Human Nutrition. The other researchers were senior men, happy to be left alone, free to do their own work. I, in contrast, was a novice, badly in need of advice and guidance; I tried to see Sinclair but after half a dozen attempts realized it was a hopeless business.

The work went badly from the start. I did not know what strength the TOCP should be, in what medium it should be given, or whether it should be sweetened to disguise its bitter taste. The worms and frogs at first refused the TOCP delicacies I concocted. The chickens, it seemed, would gobble anything—an unlovely sight. Despite their gobbling and pecking and squawking, I started to grow fond of my chickens, to take a certain pride in their noisiness and vigor, and to appreciate their distinctive behaviors and characteristics. In a few weeks, the TOCP took effect, and the chickens' legs started to weaken. At this point, thinking that TOCP might have some similarity to nerve gases (which disrupt the neurotransmitter acetylcholine), I gave anticholinergic drugs as an antidote to half of the semi-paralyzed birds. Misjudging the dose, I managed to kill them all. Meanwhile, the hens which had been spared the antidote grew weaker and weaker, a sight I could hardly bear to watch. The end, for me and for my research, was seeing my favorite hen—she had no name, but number 4304 was an animal of unusually docile and sweet disposition—sink to the ground on her paralyzed legs, chirping piteously. When I sacrificed her (using chloroform), I found she had damage to the myelin sheaths of peripheral nerves and nerve axons in the spinal cord, like the human victims who had come to autopsy.

I also found that TOCP knocked out the sudden curling reflex in earthworms, though not their other movements, that it damaged their myelinated nerve fibers but not their unmyelinated ones. But I felt that my research as a whole was a failure and that I could never hope to be a research scientist. I wrote up a colorful and rather personal account of the work and, with this, tried to dismiss the whole wretched episode from my mind.

Depressed by this and isolated because all my friends had left the university, I felt myself sinking into a state of quiet but in some ways agitated despair. I could find no relief except in physical exercise, and every evening I went for a long run on the towpath along the Isis. After running for an hour or so, I would dive in and swim and then, wet and a little chilled, run back to my mean digs opposite Christ Church. I would gobble some cold dinner (I could no longer bear to eat chicken) and then write far into the night. These writings, titled "Nightcaps," were frenzied, unsuccessful efforts to forge some sort of philosophy, some recipe for living, some reason to go on.

My tutor at Queen's, who had tried to warn me against working for Sinclair, perceived my condition (I found this surprising and reassuring; I was not sure that he even knew of my existence at this point) and voiced his concern to my parents.

Between them, they decided I needed to be extricated from Oxford and put in a friendly and supportive community with hard physical work from dawn till dusk. My parents thought that a kibbutz would fill the bill, and I too, though devoid of any religious or Zionist feeling, liked the idea. And so I departed for Ein HaShofet, an "Anglo-Saxon" kibbutz near Haifa where English would be spoken until, it was hoped, I became fluent in Hebrew.

I spent the summer of 1955 in the kibbutz. I was given a choice: I could work in the tree nursery or with chickens. I had a horror of chickens now and opted for the tree nursery. We got up before dawn, had a large communal breakfast, and then set off for our work.

I was amazed at the huge bowls of chopped liver at every meal, including breakfast. There were no cattle on the kibbutz, and I did not see how its chickens alone could provide the hundred pounds or so of chopped liver we consumed every day. When I enquired, there was laughter, and I was told that what I had taken for liver was chopped eggplant, something I had never tasted in England.

I was on good, at least conversational, terms with everybody but on close terms with nobody. The kibbutz was full of families or, rather, constituted a single super-family in which all the parents looked after all the children. I stood out as a single person with no intention of making my life in Israel (as so many of my cousins planned to do). I was not good at small talk, and in my first two months, despite intensive immersion in the *ulpan*, I learned very little Hebrew, though in my tenth week I suddenly started to understand and utter Hebrew phrases. But the hardworking physical life and the presence of friendly, thoughtful people around me served as an anodyne to the lonely, torturing months in Sinclair's lab, when I was so shut up in my own head.

And there were great physical effects, too; I had gone to the kibbutz as a pallid, unfit 250 pounds, but when I left it three months later, I had lost nearly 60 pounds and, in some deep sense, felt more at home in my own body.

After I left the kibbutz, I spent a few weeks traveling to other parts of Israel to get a feeling of the young, idealistic, beleaguered state. In the Passover service, recalling the exodus of the Jews from Egypt, we would always say, "Next year in Jerusalem," and now, finally, I saw the city where Solomon had built his temple a thousand years before Christ. But Jerusalem was divided at this time, and one could not go into the old city.

I explored other parts of Israel: the old port of Haifa, which I loved; Tel Aviv; and the copper mines, reputedly King Solomon's mines, in the Negev. I had been fascinated by what I had read of kabbalistic Judaism—especially its cosmogony— and so I made my first journey, a pilgrimage in a way, to Safed, where the great Isaac Luria had lived and taught in the sixteenth century.

And then I made for my real destination, the Red Sea. Eilat had a population of a few hundred at this time, with little more than tents and shacks (it is now a glittering seafront of hotels, with a population of fifty thousand). I snorkeled practically all day and had my first experiences of scuba diving, still relatively primitive at this time. (It had become far easier and more streamlined by the time I got my certification as a scuba diver in California a few years later.)

I wondered again, as I had wondered when I first went to

Oxford, whether I really wanted to become a doctor. I had become very interested in neurophysiology, but I also loved marine biology, especially marine invertebrates. Could one combine them, perhaps, by doing invertebrate neurophysiology, especially studying the nervous systems and behaviors of cephalopods, those geniuses among invertebrates?[2]

A part of me would have liked to stay at Eilat for the rest of my life, swimming, snorkeling, scuba diving, doing marine biology and invertebrate neurophysiology. But my parents were getting impatient; I had idled long enough in Israel; I was "cured" now; it was time to return to medicine, to start clinical work, seeing patients in London. But I had one more thing I needed to do—something which had been unthinkable before. I was twenty-two, I now thought, good-looking, tanned, lean, and still a virgin.

I had been to Amsterdam a couple of times with Eric; we loved the museums and the Concertgebouw (it was here that I first heard Benjamin Britten's *Peter Grimes*, in Dutch). We loved the canals lined with tall, stepped houses; the old Hortus Botanicus and the beautiful seventeenth-century Portuguese synagogue; the Rembrandtplein with its open-air cafés; the fresh herrings sold in the streets and eaten on the spot; and the

---

2. My examiner in zoology when I had taken the Higher School Certificate exam in 1949 was the great zoologist J. Z. Young, who had discovered the giant nerve axons of squids; it was investigation of these giant axons a few years later which led to our first real understanding of the electrical and chemical basis of nerve conduction. Young himself spent every summer in Naples, studying the behavior and brain of the octopus. I wondered whether I should try to work with him, as Stuart Sutherland, my contemporary at Oxford, was now doing.

general atmosphere of cordiality and openness which seemed peculiar to the city.

But now, fresh from the Red Sea, I decided to go to Amsterdam alone, to lose myself there—specifically, to lose my virginity. But how does one go about doing this? There are no textbooks on the subject. Perhaps I needed a drink, several drinks, to damp down my shyness, my anxieties, my frontal lobes.

There was a very pleasant bar on Warmoesstraat, near the railway station; Eric and I had often been there for a drink together. But now, by myself, I drank hard—Dutch gin for Dutch courage. I drank till the bar went in and out of focus and sounds seemed to swell and retreat. I did not realize until I stood up that I was unsteady on my feet, so unsteady that the barman said, "*Genoeg!* Enough!" and asked if I needed any help getting back to my hotel. I said no, my hotel was just across the street, and staggered out.

I must have blacked out, for when I came to the next morning, I was in not my own bed but someone else's. There was the friendly smell of coffee brewing, followed by the appearance of my host, my rescuer, in a dressing gown, with a cup of coffee in each hand.

He had seen me lying dead drunk in the gutter, he said, had taken me home . . . and buggered me. "Was it nice?" I asked.

"Yes," he answered. Very nice—he was sorry I was too out of it to enjoy it as well.

We talked more over breakfast—about my sexual fears and inhibitions and the forbidding and dangerous atmosphere in England, where homosexual activity was treated as a crime. It was quite different in Amsterdam, he said. Homosexual activity between consenting adults was accepted, not illegal, not regarded as reprehensible or pathological. There were many bars, cafés, and clubs where one could meet other gay people (I had never heard the word "gay" in this connection before). He would be happy to take me to some of them or just give me their names and whereabouts and let me fend for myself.

"There is no need," he said, suddenly getting serious, "to get dead drunk, pass out, and lie in the gutter. This is a very sad—even dangerous—thing to do. I hope you will never do it again."

I cried with relief when we spoke and felt that some huge burden, a burden above all of self-accusation, had been lifted or at least much lightened.

In 1956, after my four years in Oxford and my adventures in Israel and Holland, I moved back home and started as a medical student. In those thirty months or so, I rotated through medicine, surgery, orthopedics, pediatrics, neurology, psychiatry, dermatology, infectious diseases, and other specialties denoted only by letters—GI, GU, ENT, OB/GYN.

To my surprise (but my mother's gratification), I had a special feeling for obstetrics. In those days, babies were delivered at home (I myself had been born at home, as were all my brothers). Deliveries were largely in the hands of midwives, and we, as medical students, would assist the midwives. A phone call would come, often in the middle of the night; the hospital operator would give me a name and an address, and sometimes she would add, "Hurry!"

The midwife and I, on our bicycles, would converge on the house, go to the bedroom or occasionally the kitchen; it was sometimes easier to deliver on a kitchen table. The husband

and family would be waiting in the next room, their expectant ears tuned for the baby's first cry. It was the human drama of all this which excited me; it was real in a way that hospital work was not and our only chance to *do* something, to play a role, outside the hospital.

As medical students, we were not overloaded with lectures or formal instruction; the essential teaching was done at a patient's bedside, and the essential lesson was to listen, to get the "history of the present condition" from the patient and ask the right questions to fill in the details. We were taught to use our eyes and ears, to touch, to feel, even to smell. Listening to a heartbeat, percussing the chest, feeling the abdomen, and other forms of physical contact were no less important than listening and talking. They could establish a bond of a deep, physical sort; one's hands could themselves become therapeutic tools.

•

I qualified on December 13, 1958, and I had a couple of weeks in hand; my house job at the Middlesex was not due to start until the first of January.[3] I was excited—and amazed—to find myself a doctor, to have made it finally (I never thought I would, and sometimes even now, in my dreams, I am still stuck in an eternal studenthood). I was excited, but I was terrified too. I felt sure I would do everything wrong, make a fool of myself, be seen as an incurable, even dangerous bungler. I thought a temporary house job in the weeks before I started at the Middlesex might give me needed confidence and competence, and I managed to get such a job a few miles outside London, at a hospital

---

3. In the United States, this would have been called an internship; in England, interns are referred to as housemen, and residents as registrars.

in St. Albans where my mother had worked as an emergency surgeon during the war.

On my first night, I was called at 1:00 a.m.; a baby had been admitted with bronchiolitis. I hurried down to the ward to see my first patient—a four-month-old infant, blue around the lips, with a high fever, rapid breathing, and wheezing. Could we— the nursing sister and I—save him? Was there any hope? Sister, seeing that I was terrified, gave me the support and guidance I needed. The little boy's name was Dean Hope, and absurdly, superstitiously, we took this as a good omen, as if his very name could sweeten the Fates. We worked hard all night, and when the pale grey winter day dawned, Dean was out of danger.

•

On January 1, I started working at the Middlesex Hospital. The Middlesex had a very high reputation, even though it lacked the antiquity of "Barts"—St. Bartholomew's, a hospital dating back to the twelfth century. My older brother David had been a medical student at Barts. The Middlesex, a relative newcomer founded in 1745, was housed, in my day, in a modern building from the late 1920s. My eldest brother, Marcus, had trained at the Middlesex, and now I was following in his footsteps.

I did a six-month house job on the medical unit at the Middlesex and then another six months on the neurological unit, where my chiefs were Michael Kremer and Roger Gilliatt, a brilliant but almost comically incongruous pair.

Kremer was genial, affable, suave. He had an odd, slightly twisted smile, whether from a habitually ironical view of the world or the residue of an old Bell's palsy, I was never sure. He seemed to have all the time in the world for his housemen and his patients.

Gilliatt was much more forbidding: sharp, impatient, edgy, irritable, with (it sometimes seemed to me) a sort of suppressed fury that might explode at any moment. An undone button, we housemen felt, might provoke him to a rage. He had huge, ferocious, jet-black eyebrows—instruments of terror for us juniors. Gilliatt, recently appointed, was still in his thirties, one of the youngest consultants in England.[4] This did not diminish his formidable aspect; it might have heightened it. He had won a Military Cross for outstanding bravery in the war, and he had a rather military bearing. I was terrified of Gilliatt and became almost paralyzed with fear when he asked me a question. Many of his housemen, I would later find, had similar reactions.

Kremer and Gilliatt had very different approaches to examining patients. Gilliatt would have us go through everything methodically: cranial nerves (none to be omitted), motor system, sensory system, etc., in a fixed order, never to be deviated from. He would never leap ahead prematurely, home in on an enlarged pupil, a fasciculation, an absent abdominal reflex, or whatever.[5] The diagnostic process, for him, was the systematic following of an algorithm.

Gilliatt was preeminently a scientist, a neurophysiologist by training and temperament. He seemed to regret having to deal with patients (or housemen), though he was, I was later to learn, a completely different person—genial and supportive—when he was with his research students. His real interests,

---

4. This was indeed impressive, though my mother, I could not help thinking, had become a consultant at the age of twenty-seven.

5. Valentine Logue, their neurosurgical colleague on the ward above, used to ask junior physicians if they saw anything "wrong" about his face, and only then would we realize that there was something odd about his eyes: one of his pupils was much larger than the other. We speculated endlessly as to why this was so, but Logue never enlightened us.

his passions, were all related to the electrical investigation of peripheral nerve disorders and of muscle innervation, where he was on his way to becoming a world authority.

Kremer, on the other hand, was intuitive in the extreme; I remember him once making a diagnosis on a newly admitted patient as soon as we entered the ward. He spotted the patient thirty yards away, clutched my arm excitedly, and whispered "Jugular foramen syndrome!" in my ear. This is an exceedingly rare syndrome, and I was astonished that he could identify it, across the length of the ward, at a glance.

Kremer and Gilliatt made me think of Pascal's comparison between intuition and analysis in the opening of the *Pensées*. Kremer was preeminently intuitive; he saw everything at a glance, often more than he could put into words. Gilliatt was primarily analytic, looking at phenomena one at a time but seeing in depth the physiological antecedents or consequences of each.

Kremer's sympathy or empathy was very remarkable. He seemed to read his patients' minds, to know intuitively all their fears and hopes. He observed their movements and postures like a theater director with his actors. One of his papers—a favorite of mine—was called "Sitting, Standing, and Walking." It showed how much he observed and understood even before doing a neurological exam, even before the patient opened his mouth.

In his Friday afternoon outpatient clinics, Kremer might see thirty different patients, but each one claimed his intense, undivided, and compassionate attention. He was much loved by his patients, and all of them spoke of his kindness, of how they found his very presence therapeutic.

Kremer remained interested and often actively involved

in the lives of his housemen long after they had moved on to other jobs. He advised me about going to America, gave me some introductions, and twenty-five years later wrote me a thoughtful letter after reading *A Leg to Stand On*.[6]

I had less contact with Gilliatt—we were both, I think, quite shy—but he wrote to me when *Awakenings* came out in 1973 and invited me to call on him in Queen Square. I found him far less terrifying now, with an intellectual and emotional warmth I would never have suspected. The following year, he invited me again, to show the documentary film of my *Awakenings* patients there.

I was sad when Gilliatt died of cancer—he was still relatively young and very productive when it hit him—and when Kremer, who was so sociable and so loved conversation and who continued to see patients long after his "retirement," was rendered aphasic by a stroke. Both influenced me, in good but very different ways: Kremer to be more observant and intui-

---

6. Kremer wrote:

   I was asked to see a puzzling patient on the cardiology ward. He had atrial fibrillation and had thrown off a large embolus giving him a left hemiplegia, and I was asked to see him because he constantly fell out of bed at night for which the cardiologists could find no reason.

   When I asked him what happened at night he said quite openly that when he woke up in the night he always found that there was a dead, cold, hairy leg in bed with him which he could not understand but could not tolerate and he, therefore, with his good arm and leg pushed it out of bed and naturally, of course, the rest of him followed.

   He was such an excellent example of this complete loss of awareness of his hemiplegic limb but, interestingly enough, I could not get him to tell me whether his own leg on that side was in bed with him because he was so caught up with the unpleasant foreign leg that was there.

   I quoted this passage of Kremer's letter when I had occasion to describe a similar case ("The Man Who Fell out of Bed") in *The Man Who Mistook His Wife for a Hat*.

tive; Gilliatt to think always of the physiological mechanisms involved. More than fifty years later, I remember them both with affection and gratitude.

My pre-med studies in anatomy and physiology at Oxford had not prepared me in the least for real medicine. Seeing patients, listening to them, trying to enter (or at least imagine) their experiences and predicaments, feeling concerned for them, taking responsibility for them, was quite new to me. Patients were real, often passionate individuals with real problems—and sometimes choices—of an often agonizing sort. It was not just a question of diagnosis and treatment; much graver questions could present themselves—questions about the quality of life and whether life was even worth living in some circumstances.

This hit me very hard when I was a houseman at the Middlesex and Joshua, a young man and fellow swimmer, was admitted to the medical unit with odd and puzzling pains in his legs. A tentative diagnosis was made from some blood tests, and, pending further tests, he was allowed to spend the weekend at home. On Saturday night, while he was enjoying himself at a party with a crowd of young people, including some medical students, one of the students asked Joshua why he had been admitted to hospital. He said he didn't know but had been given some pills. He showed the bottle to his questioner, who, seeing "6MP" (6-mercaptopurine) on the label, blurted out, "Christ, you must have acute leukemia."

When Joshua came back from his weekend leave, he was in a desperate state of mind. He asked if the diagnosis was certain, what treatments could do, what lay in store for him. A bone marrow test was done, confirming the diagnosis, and he was told that while medication might give him a little more time, he would go downhill and die within a year, perhaps sooner.

That afternoon, I saw Joshua climbing over the railings of the balcony; our ward was on the second floor. I rushed over to the railings and pulled him back, saying whatever I could about life being worth living, even with such a condition. Reluctantly—the moment of decision had passed—Joshua let himself be persuaded back into the ward.

The strange pains quickly grew more severe and began to affect his arms and trunk as well as his legs. It became apparent that these were caused by leukemic infiltrations of the sensory nerves as they entered the spinal cord. Pain medication was of no use, though he was put on ever-stronger opiates, by mouth and injection, and finally heroin. He started to scream with pain, day and night, and at this juncture the only recourse was to give him nitrous oxide. As soon as he came round from the anesthesia, he started screaming again.

"You shouldn't have pulled me back," he said to me. "But I guess you had to." He died, racked with pain, a few days later.

It was not easy, or safe, to be an open or practicing homosexual in the London of the 1950s; homosexual activities, if detected, could lead to harsh penalties, imprisonment, or, as in Alan Turing's case, chemical castration by the mandatory administration of estrogen. Public attitudes were, on the whole, as condemnatory as the law. It was not easy for homosexuals to meet; there were some gay clubs and gay pubs, but these were constantly watched and raided by the police. There were agents provocateurs everywhere, especially in public parks and public lavatories, trained to seduce the unwary or ingenuous and bring them to destruction.

Though I made visits to "open" cities like Amsterdam as often as I could, I dared not seek any sexual partners in London, the more so as I was living at home, under my parents' vigilant eyes.

But in 1959, when I was doing house jobs in medicine and neurology at the Middlesex, I had only to walk down Charlotte Street and cross Oxford Street, and I was in Soho Square. A little farther—down Frith Street—and I would find myself in Old Compton Street, where everything was for rent or sale. Here, at Coleman's, I could get my favorite Havana cigars; a Bolivar "torpedo" could last a whole evening, and I would treat myself to one on special, festive occasions. There was a delicatessen which sold a poppy seed cake of a succulence, a lusciousness, I have never tasted since, and there was a little newspaper and confectionery shop with sexual advertisements posted in its windows, advertisements discreetly ambiguous— anything else would have been dangerous—but unmistakable in meaning.

One such was from a young man who said that he loved motorbikes and motorbike gear. He gave his first name, at least *a* name, Bud, and a phone number. I dared not linger, much less make a note of his advertisement, but my then-photographic memory absorbed it in an instant. I had never answered an advertisement, or thought of doing so, in my life, but now, after nearly a year of abstinence—I had not been to Amsterdam since the previous December—I decided to phone the enig- matic "Bud."

We chatted on the phone, circumspectly, talking mostly about our motorbikes. Bud had a BSA Gold Star, a big 500 cc single with dropped handlebars, and I had my 600 cc Norton Dominator. We arranged to meet at a motorbike café and go for a ride from there. We would recognize each other by our bikes and our kit: leather jackets, leather pants, leather boots and gloves.

We met, shook hands, admired each other's bikes, and then took off for a ride around south London. Born and bred in north-west London, I was lost in south London, so Bud led the way. I thought he looked very gallant, a knight of the roads, riding his motor-steed, clad in black leather.

We went back to his apartment in Putney for dinner—a rather bare apartment, with very few books but a lot of motorbike magazines and motorbike gear. There were photographs of motorcycles and motorcyclists all over the walls and (this I had not expected) some beautiful underwater photographs which Bud had taken; his other passion, besides motorcycling, was scuba diving. I had been initiated into scuba diving while I was at the Red Sea in 1956, and so here was another avocation we shared (quite an unusual one in the 1950s). Bud had a lot of diving gear, too; this was before the days of wet suits and neoprene, and one wore dry suits of heavy rubber.

We had a beer and then, quite suddenly, Bud said, "Let's go to bed."

We made no attempt to find out more about each other; I knew nothing about Bud, his work, or even his last name, and he knew as little about me, but we knew (intuitively, unerringly) what we wanted, how we could pleasure ourselves and each other.

There was no need to say afterwards how much we had

enjoyed our encounter, how much we wanted to meet again. I was about to go to Birmingham for a six-month house job in surgery, but this was a problem easily dealt with. I could shoot down to London on my bike on Saturdays to spend the night at home with my parents, but I could arrive early and spend the afternoon with Bud first, and the following morning we could go for a ride together. I loved our rides on crisp Sunday mornings, especially if I stowed my own bike and rode pillion with Bud, so closely jammed together we sometimes felt like a single leather animal.

I was in a state of uncertainty about my own future at this time: my house jobs would come to an end in June of 1960, and I would then be eligible for the draft (my call-up had been postponed through the years of studenthood and internship).

I said nothing while I was ruminating on this, but in June I wrote to Bud, saying that I would be leaving England for Canada on my birthday, July 9, and might not return. I did not think this would affect him very much; we had been motorcycle buddies and bed buddies but, I thought, nothing more. We never spoke about any feelings for each other. But Bud sent me a passionate, painful letter in reply; he felt desolate, he said, had sobbed when he received my letter. I felt stricken when I received his letter and realized, too late, that he must have fallen in love with me and that now I had broken his heart.

Leaving the Nest

A s a boy, reading novels by Fenimore Cooper and watching cowboy films, I had formed a romantic view of America and Canada. The rugged open spaces of the American West portrayed in John Muir's books and Ansel Adams's photographs seemed to promise an openness and freedom and an ease which England, still recovering from the war, lacked.

As a medical student in England, my military service had been deferred, but as soon as I finished my house jobs, I was due to report for duty. I did not like the idea of military service much (although my brother Marcus had enjoyed it, and his knowledge of Arabic got him posted to Tunisia, Cyrenaica, and North Africa). I had signed up for an attractive alternative, doing a three-year stint as a doctor in the Colonial Service, and I had elected to do my three years in New Guinea. But the Colonial Service itself was shrinking, and its medical service option came to an end just before I finished medical school. Obligatory military service itself was due to end just months after my enlistment date in August.

The feeling that an attractive, exotic posting in the Colonial Service was no longer possible and the fact that I would be one of the very last conscripts infuriated me and was another factor in making me leave England. Nonetheless, I felt that I had, in some sense, a moral duty to serve. These conflicting feelings drove me, when I arrived in Canada, to volunteer for the Royal Canadian Air Force. (I was entranced by a line of

Auden's, about the airman's "laughter in leather.") Service in Canada, a Commonwealth country, would have been accepted as equivalent to military service in England, an important consideration if I returned to England.

There were other reasons to escape England, as my brother Marcus had done ten years earlier when he went to live in Australia. Large numbers of highly qualified men and women left England in the 1950s (in the so-called brain drain), for professions and universities were crowded in England, and (as I saw when I did my neurology internship in London) brilliant and accomplished people could get stuck for years in subservient roles, never enjoying autonomy or responsibility. I thought that America, with a medical system far more capacious and far less rigid than that in England, would have room for me. I was also moved, as Marcus had been, by a feeling that there were too many Dr. Sackses in London: my mother, my father, my older brother David, an uncle, and three first cousins, all competing for space in London's crowded medical world.

I flew to Montreal on July 9, my twenty-seventh birthday. I spent a few days there, staying with relatives, visiting the Montreal Neurological Institute, and making contact with the RCAF. I wanted to become a pilot, I told them, but after some tests and interviews, they said that with my background in physiology I would be best placed in research. A very high-up officer, a Dr. Taylor, interviewed me at length and invited me to spend an evaluatory weekend with him. At the end of this, sensing my ambivalence, he said, "You are clearly talented, and we would love to have you, but I am not sure about your motives for joining. Why don't you take three months to travel, think about things. If you still want to join up then, contact me."

I was relieved by this; I had a sudden sense of freedom and lightheartedness and decided to make the most of my three-month "leave."

I set out to travel across Canada and, as always when traveling, kept a journal. I wrote only brief letters to my parents while I was moving across Canada, and I did not get a chance to write to them more fully until I got to Vancouver Island. There I composed an enormous letter to them, detailing my travels.

Trying to summon a picture of Calgary, of the Wild West, for my parents, I let my imagination loose; I doubt if the real Calgary was quite as exotic as I portrayed:

Calgary has just finished its annual "stampede," and the streets are full of loafing cowboys in jeans and buckskins, sitting the long days out with their hats crushed over their faces. But Calgary also has 300,000 citizens. It is a boom town. Oil has brought a huge influx of prospectors, investors, engineers to it. The Old West life has been overwhelmed by refineries, and factories, and by offices and skyscrapers. . . . There are also tremendous fields of uranium ore, gold and silver, and the base metals, and you can see little packets of gold dust passed from hand to hand in the taverns, and men made of solid gold behind their tanned faces and filthy overalls.

Then I went back to the joys of traveling:

I took the Canadian Pacific Railroad to Banff, roaming excitedly in the train's "scenic dome." We passed from the boundless flat prairies through the low spruce-covered foothills of the Rockies, climbing gently all the time. And gradually the air became cooler, and the scale of the country more vertical. The hillocks grew to hills, and the hills to mountains, higher and jaggeder with each mile we progressed. We puffed punily in the floor of a valley, and snowcapped mountains soared tremendous about us. The air was so clear that one could see peaks a hundred miles away, and the mountains beside us seemed to be rearing over our very heads.

From Banff, I went deeper into the heart of the Canadian Rockies. I kept a particularly detailed journal here, and later I reworked it for a piece I called "Canada: Pause, 1960."

CANADA: PAUSE, 1960

How I have moved! I have traveled nearly three thousand miles in less than two weeks.

Now there is stillness—such a stillness as I have never heard before in all my life. Soon I shall start moving again, and perhaps I will never stop.

I am lying in a high alpine meadow, more than eight thousand feet above sea level. Yesterday I wandered near our lodge with three botanical ladies from Calgary, lean and tough as Amazons they were, and learned from them the names of many flowers.

The meadow is dominated by mountain avons, in seed now, like huge dandelion heads, alight and floating as they catch the morning sun. Indian paintbrush, from a faint cream to an intense vermillion. Chalice-cup, globe flowers, valerians, and saxifrages; contorted lousewort and stinking fleabane (two of the loveliest, despite their names), Arctic raspberries and strawberries, which rarely fruit; the three-leaved strawberries catch and hold at their center a flashing drop of dew. Heart-shaped arnicas, calypso orchids, cinquefoils, and columbines. Glacial lilies and alpine speedwell. Some of the rocks are covered by brilliant lichens, blazing in the distance like great masses of precious stones; others are clustered with succulent stonecrops, which burst lasciviously under the pressing finger.

We are far above the territory of lofty trees. There are many shrubs—willow and juniper, bilberry and buffalo berry—but above the timberline only the larches, with their chaste white trunks and downy foliage.

There are gophers, picas, squirrels, and chipmunks, sometimes a marmot in the shadow of a rock. Magpies, warblers, wrens, and thrushes. Bears galore, black and brown, though grizzlies are rare. Elk and moose in the lower pastures. I have seen an enormous shadow wing across the sun, and known at once it was a Rockies eagle.

Higher and higher—all life dying away, everything becoming a uniform gray, till mosses and lichens are the lords of creation once more.

Yesterday I joined the Professor, his family, and friend, "Old Marshall," whom he called "brother," and they looked like brothers, but were only friends and colleagues. I rode with them into a vast mountain plateau so high that we could look down upon the massed cumulus about us.

"Man has made no changes here!" cried the Professor, "he has only enlarged the goat trails." I have no words for that feeling, nor had I ever had it before, which comes from the knowledge that one is far away from all humanity, alone in a thousand square miles. We rode in silence, for speech would have been absurd. It seemed the very summit of the world. Later we descended, our horses treading delicately among the undergrowth, to the glacial string of lakes with their strange names—Lake Sphinx, Lake Scarab, and Lake Egypt. Ignoring their cautious warnings, I stripped off my sweaty clothes and dived into the clear waters of Egypt, and floated on my back. To one side rose the Pharaoh Mountains, their old faces marked with gigantic hieroglyphics; but the other peaks were all unnamed—they may well remain so.

Coming back we passed a great glacial basin full of smooth moraine.

"Think!" cried the Professor. "This prodigious bowl was filled with ice to a depth of three hundred feet. And when we and our children are dead, seeds will have sprouted in the silt, and a young forest will nod over these stones. Here before you is one scene of a geological drama, past and future implicit in the present you perceive, and all within the span of a single human generation, and a human memory."

I glanced at the Professor as he stood there, a tiny figure against the seven-hundred-foot wall of rock and ice; absurd in his battered hat and trousers, yet full of dignity and command. One saw the might of the glaciers and the torrents, and they were as nothing to the might of this proud insect who surveyed and understood them.

The Professor was a wonderful companion. On a strictly practical level, he taught me to recognize glacial cirques and the different species of moraine; to decipher the trails of moose and bear, and the ravages of porcupines; to survey the terrain closely for marshy or treacherous ground; to fix landmarks in my mind, so that I could never get lost; to mark the sinister lens-shaped clouds which portend freak storms. But his range was enormous, perhaps complete. He spoke of law and sociology, of economics; of politics and business, advertising; of medicine, psychology, and mathematics.

I had never known a man so profoundly in touch with every aspect of his environment—physical, social, human; yet he was enriched by a mocking insight of his own mind and motives which balanced, and rendered personal, everything he said.

I had met the Professor the evening before, and confided to

him the story of my flight from family and country, and my hesitations about continuing medicine.

"My chosen profession!" I exclaimed bitterly. "Others chose it for me. Now I want only to wander and write. I think I shall be a logger for a year."

"Forget it!" said the Professor, shortly. "You'd be wasting your time. Go and see medical schools, universities, in the States. The States is for you. Nobody'll push you around. If you're good, you go up. If you're phony, they soon catch on.

"Travel now by all means—if you have the time. But travel the right way, the way I travel. I am always reading and thinking of the history and geography of a place. I see its people in terms of these, placed in the social framework of time and space. Take the prairies, for example; you're wasting your time visiting these unless you know the saga of the homesteaders, the influence of law and religion at different times, the economic problems, the difficulties of communication, and the effects of successive mineral finds.

"Forget about lumber camps. Go to California. See the redwoods. See the missions. See Yosemite. See Palomar—it's a supreme experience for an intelligent man. I once talked to Hubble and found he knew a prodigious amount of law. Did you know that he was a lawyer before he turned to the stars? And go to San Francisco! It's one of the twelve most interesting cities in the world. California has immense contrasts— the utmost wealth and the most hideous squalor. But there's beauty and interest everywhere.

"I have crossed America every way more than a hundred times. I have seen everything. I'll tell you where to go if you tell me what you want. Well, what have you got to say?"

"I have run out of money!"

"I shall lend you whatever you need, and you can repay me when you like."

The Professor had known me then a single hour.

The Professor and Marshall love the Rockies, and come to them every summer, as they have done for twenty years. On our return from Lake Egypt, they took me off the trail and deep into the forest until we came to a low dark cabin, half buried in the ground. The Professor delivered a brief illuminating lecture:

"This is Bill Peyto's cabin. Only three people in the world besides us know where it is; it is officially listed as having been destroyed by fire. Peyto was a nomad and a misanthrope, a great hunter and observer of wildlife, and the father of uncountable bastards. He has a lake and a mountain named after him. Some slow malady attacked him in 1926, and finally he could live by himself no longer. He rode down into Banff, a wild and legendary stranger whom everyone knew of but nobody had seen. He died there soon after."

I advanced towards the darkened and rotting hut. Its door was askew, and upon it I deciphered a faint scrawl: BACK IN AN HOUR. Inside I saw his cooking utensils and ancient preserves, his mineral specimens (he operated a small talc mine), fragments of his journal, and the *Illustrated London News* from 1890 to 1926. A temporal cross-section of a man's life, cut clean by circumstance. I thought of the *Marie Celeste*. It is evening now, and I have spent the whole day lying in this broad meadow, chewing a blade of grass, and looking at the mountains and the sky. I have reflected, and I have nearly filled my notebook.

On a summer evening, at home, the setting sun is light-

ing up the hollyhocks and the cricket stumps stuck in the
back lawn. Today is a Friday, and this means my mother will
light the Sabbath candles, murmuring as she cups the flames
a silent prayer whose words I never knew. My father will
don a little cap and, lifting the wine, will praise God for his
fecundity.

A little wind has sprung up, breaking at last the long stillness of the day, giving a restless quiver to the grass and flowers. It is time to get up and move, away from here, and onto
the road again. Did I not promise myself that soon I would be
in California?

•

Having traveled by plane and train, I decided to complete
my westward journey by hitchhiking—and almost immediately got conscripted for firefighting. I wrote to my parents,

British Columbia has had no rain for more than thirty days, and
there are forest fires raging everywhere (you have probably read
about them). A sort of martial law exists, and the forest commission can conscript anyone they feel is suitable. I was quite
glad of the experience, and spent a day in the forests with other
bewildered conscripts, dragging hoses to and fro, and trying to
be useful. However it was only for one fire they wanted me, and
when at last we shared a beer over its smoking dwindling ruin, I
felt a real glow of confraternal pride that it had been vanquished.

British Columbia at this time of the year seems bewitched.
The sky is low and purple, even at midday, from the smoke of
innumerable fires, and the air has a terrible stultifying heat and
stillness. People seem to move and crawl with the tedium of a
slow motion film, and a sense of imminence is never absent.

In all the churches prayers are said for rain, and god knows what strange rites are practiced in private to make it come. Every night lightning will strike somewhere, and more acres of valuable timber conflagrate like tinder. Or sometimes there is just an instantaneous apparently sourceless combustion arising like some multifocal cancer in a doomed area.

Not wanting to be re-conscripted for firefighting—I enjoyed a day of it, but that was enough—I took a Greyhound bus for the remaining six hundred miles to Vancouver.

From Vancouver, I took a boat to Vancouver Island and ensconced myself in a guesthouse in Qualicum Beach (I liked the name Qualicum because it brought to mind Thudichum, the nineteenth-century biochemist, and *Colchicum,* the autumn crocus). Here I allowed myself a few days' rest from traveling and composed an eight-thousand-word letter to my parents, ending with the here and now:

The Pacific Ocean is warm (about 75°) and enervating after the glacial lakes. I went fishing today with an ophthalmologist here, fellow called North, once at Marys and the National, now in practice in Victoria. He calls Vancouver Island a "little bit of heaven which got left somehow," and I think he's right in a way. It has forests and mountains and streams and lakes and the ocean. . . . By the way, I caught six salmon, one just lets the line trail, and they bite, bite; sweet silvery beauties, which I shall have for breakfast tomorrow.

"I will descend to California in two or three days," I added, "probably by Greyhound bus, as I gather they are particularly hard on hitchhikers, and sometimes shoot them on sight."

•

I arrived in San Francisco on a Saturday evening, and that night I was taken to dinner by friends I had met in London. The next morning they picked me up, and we drove over the Golden Gate Bridge, up the piney flanks of Mount Tamalpais to the cathedral-like calm of Muir Woods. I became silent with awe beneath the redwoods, and it was at that moment that I decided I wanted to stay in San Francisco, with its wonderful environs, for the rest of my life.

There were innumerable things to do: I had to get a green card; I had to find a place to work, a hospital which would employ me, informally and without pay, during the months it would take to get my green card; I wanted all my things from England—clothes, books, papers, and (not least) my faithful Norton motorbike; I needed all sorts of documents; and I needed money.

I could be lyrical and poetic when I wrote to my parents, but now I had to be practical and pragmatic. I had ended my giant letter from Qualicum Beach by thanking my parents:

> IF I STAY in Canada, I will have a reasonably generous salary and time off. I should be able to save, and even to return something of the money which you have lavished on my life for twenty-seven years. As for the other intangible and incalculable things you have given me, I can only repay these by leading a fairly happy and useful life, keeping in touch with you, and seeing you when I can.

Now, only a week later, everything had changed. I was no longer in Canada, no longer thinking of life in the RCAF, no longer thinking of returning to England. I wrote to my parents again—fearfully, guiltily, but resolutely—telling them of my decision. I imagined their rage, their reproaches at my decision; had I not abruptly (and perhaps deceitfully) taken off and turned my back on them, on all my friends and family, on England itself?

They responded nobly, but they also expressed their sadness at our separation, in words which tear at me as I read them fifty years later—words which must have been wrenched out of my mother, for she rarely spoke of her feelings.

August 13, 1960

My dear Oliver,

Many thanks for your various letters and cards. I have read them all—with pride at your literary prowess, happiness that you are enjoying your holiday, but with a big element of grief and sadness at the thought of your prolonged absence. When you were born, people congratulated us on what they considered a wonderful family of four sons! Where are you all now? I feel lonely and bereft. Ghosts inhabit this house. When I go into the various rooms I feel overcome with a sense of loss.

My father, in a different mode, wrote, "We are quite reconciled to a comparatively empty house at Mapesbury." But then he added a postscript:

When I say that we are reconciled to an empty house, this is of course, a half-truth. I need hardly say that we miss you very much at all times. We miss your cheery presence, your ravenous attacks on the "fridge" and larder, your piano playing, your disporting yourself naked in your room weightlifting—your unexpected descents at midnight with your Norton. These and a host of other memories of your vital personality will always remain with us. When we contemplate this large empty house, we feel a wrench at our heart and a deep sense of loss. We realize nevertheless that you have to make your way in the world, and with you must rest the ultimate decision!

My father had written about "an empty house," and my mother wrote, "Where are you all now? . . . Ghosts inhabit this house."

But there was still a very real, substantial presence in the house, and this was my brother Michael. Michael had, in some sense, been the "odd" son from his earliest years. There always seemed something different about him; he found it difficult to make contact, he had no friends, he seemed very much to live in a world of his own.

Our elder brother Marcus's favorite world, from an early age, was one of languages; he spoke half a dozen by the age of sixteen. David's was one of music; he could have been a professional musician. Mine was one of science. But what sort of world Michael lived in, none of us knew. And yet he was very intelligent; he read continually, had a prodigious memory, and seemed to turn to books, rather than "reality," to get his knowledge of the world. My mother's eldest sister, Auntie Annie, who headed a school in Jerusalem for forty years, thought Michael so extraordinary that she left her entire library to him, even though her last sight of him had been in 1939, when he was only eleven years old.

Michael and I were evacuated together, at the start of the war, and spent eighteen months at Braefield, a hideous board-

ing school in the Midlands run by a sadistic headmaster whose
chief pleasure in life seemed to be beating the bottoms of the
little boys under his control.[1] (It was then that Michael learned
Nicholas Nickleby and *David Copperfield* by heart, though
he never explicitly compared our school to Dotheboys or our
headmaster to Dickens's monstrous Mr. Creakle.)

In 1941, Michael, now thirteen, went on to another boarding
school, Clifton College, where he was unmercifully bullied. In
Uncle Tungsten, I wrote about how Michael's first psychosis
developed:

> My Auntie Len, who was staying with us, spied Michael as he
> came, half naked, from the bath. "Look at his back!" she said to
> my parents, "it's full of bruises and weals! If this is happening
> to his body," she continued, "what is happening to his mind?"
> My parents seemed surprised, said they had noticed nothing
> amiss, that they thought Michael was enjoying school, had no
> problems and was "fine."
>
> Soon after this, when he was fifteen, Michael became psy-
> chotic. He felt a magical and malignant world was closing about
> him. He came to believe that he was "the darling of a flagello-
> maniac God," as he put it, subject to the special attentions of "a
> sadistic Providence." Messianic fantasies or delusions appeared
> at the same time—if he was being tortured or chastised, this was
> because he was (or might be) the Messiah, the one for whom we
> had waited so long. Torn between bliss and torment, fantasy
> and reality, feeling he was going mad (or perhaps so already),
> Michael could no longer sleep or rest, but agitatedly strode to

---

1. I have written of the school, and its effects on us, in much more detail
in *Uncle Tungsten*.

and fro in the house, stamping his feet, glaring, hallucinating, shouting.

I became terrified of him, for him, of the nightmare which was becoming reality for him. What would happen to Michael, and would something similar happen to me, too? It was at this time that I set up my own lab in the house, and closed the doors, closed my ears, against Michael's madness. It was not that I was indifferent to Michael; I felt a passionate sympathy for him, I half-knew what he was going through, but I had to keep a distance also, create my own world of science so that I would not be swept into the chaos, the madness, the seduction, of his.

The effect of this on my parents was devastating; they felt alarm, pity, horror, and above all bewilderment. They had a word for it—"schizophrenia"—but why should it have singled out Michael and at such an early age? Was it the terrible bullying at Clifton? Was it something in his genes? He had never seemed a normal child; he was awkward, anxious, perhaps "schizoid" even before his psychosis. Or—the most painful for my parents to consider—was it the result of the way they had treated or mistreated him? Whatever it was—nature or nurture, bad chemistry or bad bringing up—medicine could surely come to his aid. At sixteen, Michael was admitted to a psychiatric hospital and given twelve "treatments" of insulin shock therapy; this entailed bringing his blood sugar down so low that he lost consciousness and then restoring it with a glucose drip. This was the first line of treatment for schizophrenia in 1944, to be followed, if need be, by electroconvulsive treatment or lobotomy. The discovery of tranquilizers was still eight years in the future.

Whether as a result of the insulin comas or a natural process of resolution, Michael returned from the hospital three months later, no longer psychotic but deeply shaken, feeling he might never hope to lead a normal life. He had read Eugen Bleuler's *Dementia Praecox; or, The Group of Schizophrenias* while in hospital.

Marcus and David had enjoyed a day school in Hampstead, a few minutes' walk from our house, and Michael was glad now to continue his education there. If he was changed by his psychosis, this was not immediately apparent; my parents chose to think of it as a "medical" problem, something from which one could make a complete recovery. Michael, however, saw his psychosis in quite different terms; he felt it had opened his eyes to things that he had never previously thought about, in particular the downtroddenness and exploitation of the world's workers. He started reading a communist newspaper, *The Daily Worker*, and going to a communist bookshop in Red Lion Square. He devoured Marx and Engels and saw them as the prophets, if not the messiahs, of a new world era.

By the time Michael was seventeen, Marcus and David had finished medical school. Michael did not want to become a doctor, and he had had enough of school. He wanted to *work*— were workers not the salt of the earth? One of my father's patients had a large accounting business in London and said he would be happy to have Michael as an accounting apprentice or in any other capacity he wanted. Michael was quite clear about the role he would like to have; he wanted to be a messenger, to deliver letters or packages that were too important and urgent to be left to the post. In this, he was absolutely meticulous; he would insist on putting whatever message or packet he was entrusted with directly into the hands of its designated recipient and no one else. He loved walking around

London and spending his lunch hours on a park bench, if the weather was nice, reading *The Daily Worker*. He once told me that the seemingly humdrum messages he delivered might have hidden, secret meanings, apparent only to the designated recipient; this was why they could not be entrusted to anyone else. Though he might *appear* to be an ordinary messenger with ordinary messages, Michael said, this was by no means the case. He never said this to anyone else—he knew it would sound bizarre, if not mad—and he had begun to think of our parents, his older brothers, and the entire medical profession as determined to devalue or "medicalize" everything he thought and did, especially if it had any hint of mysticism, for they would see it as an intimation of psychosis. But I was still his little brother, just twelve years old, not yet a medicalizer, and able to listen sensitively and sympathetically to anything he said, even if I could not fully understand it.

Every so often—it happened many times in the 1940s and early 1950s, while I was still at school—he became floridly psychotic and delusional. Sometimes there was warning of this: he would not *say*, "I need help," but he would indicate it by an extravagant act, such as flinging a cushion or an ashtray to the floor in his psychiatrist's office (he had been seeing one since his initial psychosis). This meant, and was understood to mean, "I'm getting out of control—take me into hospital."

At other times, he gave no warning but would get into a violently agitated, shouting, stamping, hallucinated state—on one occasion, he hurled my mother's beautiful old grandfather clock against a wall—and at such times my parents and I would be terrified of him. Terrified, and deeply embarrassed—how could we invite friends, relatives, colleagues, *anyone*, to the house with Michael raving and rampaging upstairs? And what would their patients think? Both my parents had their medical offices in the house. Marcus and David also felt reluctant about inviting their friends into (what sometimes seemed to be) a madhouse. A sense of shame, of stigma, of secrecy, entered our lives, compounding the actuality of Michael's condition.

I found it a great relief when I took weekends or holidays away from London—holidays which, besides everything else, were holidays from Michael, from his sometimes intolerable presence. And yet there were other times when his native sweetness of character, his affectionateness, his sense of humor, shone out again. At such times, one realized, even when he was raving, that the real Michael, sane and gentle, was there underneath his schizophrenia.

•

When, in 1951, my mother learned of my homosexuality and said, "I wish you had never been born," she was speaking, though I am not sure I realized this at the time, out of anguish as much as accusation—the anguish of a mother who, feeling she had lost one son to schizophrenia, now feared she was losing another son to homosexuality, a "condition" which was regarded then as shameful and stigmatizing and with a deep power to mark and spoil a life. I was her favorite son, her "mugwump" and "pet lamb," when I was a child, and now I was "one of *those*"—a cruel burden on top of Michael's schizophrenia.

The situation changed for Michael and for millions of other schizophrenics, for better and worse, around 1953, when the first tranquilizer—a drug called Largactil in England, Thorazine in the United States—became available. The tranquilizers could damp down and perhaps prevent the hal-

lucinations and delusions, the "positive symptoms" of schizophrenia, but this could come at great cost to the individual. I
first saw this, shockingly, in 1956 when I came back to London
after my months in Israel and Holland and saw that Michael
was bent over and walked with a shuffling gait.

"He's grossly parkinsonian!" I said to my parents.

"Yes," they said, "but he's much calmer on the Largactil.
He's gone a year without a psychosis." I had to wonder, however, how Michael felt. He was distressed at the parkinsonian
symptoms—he had been a great walker, a strider, before—but
even more upset by the mental effects of the drug.

He was able to continue at his job, but he had lost the mystical feeling that gave depth and meaning to his messengering;
he had lost the sharpness and clarity with which he had previously perceived the world; everything seemed "muffled" now.
"It is like being softly killed," he concluded.[2]

When Michael's dose of Largactil was reduced, his parkinsonian symptoms subsided, and, more important, he felt more
alive and regained some of his mystical sensibilities—only to
explode, a few weeks later, into florid psychosis again.

In 1957, by now a medical student myself, interested in
brain and mind, I phoned Michael's psychiatrist and asked if
we could meet. Dr. N. was a decent, sensitive man who had
known Michael since his initial psychosis nearly fourteen
years earlier, and he too was disturbed by the new, drug-related
problems he was encountering with many of his patients on

_____

2. Years later, when I worked at Bronx State Hospital, I was to see gross
motor disorders, and hear similar mental complaints, from hundreds of
schizophrenic people who had been slugged with heavy doses of drugs like
Thorazine or a then-new category of drugs called butyrophenones, such as
haloperidol.

Largactil. He was trying to titrate the drug, to find a dosage which would be just enough but not too much or too little. He was, he confessed, not entirely hopeful here.

I wondered whether systems in the brain concerned with the perception (or projection) of meaning, significance, and intentionality, systems underlying a sense of wonder and mysteriousness, systems for appreciation of the beauty of art and science, had lost their balance in schizophrenia, producing a mental world overcharged with intense emotion and distortions of reality. These systems had lost their middle ground, it seemed, so that any attempt to titrate them, damp them down, could tip the person from a pathologically heightened state to one of great dullness, a sort of mental death.

Michael's lack of social skills and of ordinary day-to-day aptitudes (he could scarcely make a cup of tea for himself) demanded a social and "existential" approach. Tranquilizers have little or no effect on the "negative" symptoms of schizophrenia—withdrawal, flattening of affect, etc.—which, in their insidious, chronic way, can be more debilitating, more undermining of life, than any positive symptoms. It is a question of not just medication but the whole business of living a meaningful and enjoyable life—with support systems, community, self-respect, and being respected by others—which has to be addressed. Michael's problems were not purely "medical."

·

I could, I should, have been more loving, more supportive, while I was back in London, in medical school; I could have gone out with Michael to restaurants, films, theaters, concerts (which he never did by himself); I could have gone with him on visits to the seaside or the countryside. But I didn't, and the shame of this—the feeling that I was a bad brother, not avail-

able to him when he was in such need—is still hot within me
sixty years later.

I don't know how Michael would have responded had I
shown more initiative. He had his own severely controlled and
limited life and disliked any departure from it.

His life, now that he was on tranquilizers, was less turbulent but, it seemed to me, increasingly impoverished and con-
stricted. He no longer read *The Daily Worker*, no longer visited
the bookshop in Red Lion Square. He had once had a certain
feeling of belonging to a collective, sharing a Marxist perspective with others, but now, as his ardor cooled, he felt increasingly solitary, alone. My father hoped that our synagogue might
provide moral and pastoral support, a sense of community, for
Michael. He had been quite religious as a youth—after his bar
mitzvah, he wore tzitzit and laid tefillin daily and went to shul
whenever he could—but here too his ardor had cooled. He lost
interest in the synagogue, and the synagogue, with its diminishing community—more and more of London's Jews were
emigrating or assimilating and intermarrying in the general
population—lost interest in him.

Michael's general reading, once so intense and omnivorous—
had not Auntie Annie left her entire library to him?—dwindled
dramatically; he ceased to read books almost completely and
looked at newspapers only desultorily.

I think that despite, or perhaps because of, the tranquilizers, he had been sinking into a state of hopelessness and apa-
thy. In 1960, when R. D. Laing published his brilliant book
The Divided Self, Michael had a brief resurgence of hope. Here
was a physician, a psychiatrist, seeing schizophrenia not as a
disease so much as a whole, even privileged mode of being.
Although Michael himself sometimes called the rest of us,

the non-schizophrenic world, "rottenly normal" (great rage was embodied in this incisive phrase), he soon tired of Laing's "romanticism," as he called it, and came to regard him as a slightly dangerous fool.

When I left England on my twenty-seventh birthday, it was, among many other reasons, partly to get away from my tragic, hopeless, mismanaged brother. But perhaps, in another sense, it would become an attempt to explore schizophrenia and allied brain-mind disorders in my own patients and in my own way.

San Francisco

I had arrived in San Francisco, a city I had dreamed of for years, but I had no green card, and therefore I could not be legally employed or earn any money. I had kept in touch with Michael Kremer, my neurology chief at the Middlesex (he had been all in favor of my skipping the draft; "a complete waste of time now," he said), and when I mentioned my thoughts of going to San Francisco, he suggested that I look up his colleagues Grant Levin and Bert Feinstein, neurosurgeons at Mount Zion Hospital. They were pioneers in the art of stereotactic surgery, a technique that allowed one to insert a needle directly and safely into small, otherwise inaccessible areas of the brain.[1]

Kremer had written to introduce me, and when I met Levin and Feinstein, they agreed to take me on informally. They suggested that I help evaluate their patients before and after surgery; they could not give me a salary, since I had no green card, but they kept me going with $20 notes. (Twenty dollars was a lot at the time; an average motel cost about $3 a night, and pennies were still used in some parking meters.)

Levin and Feinstein said they would find me a room to live

---

1. It had been found that if one made small lesions in certain areas (by injecting alcohol or freezing), such lesions, far from harming patients, could break a circuit which had become hyperactive and was responsible for many symptoms of parkinsonism. Such stereotactic surgery almost ceased to be done with the coming of L-dopa in 1967 but is now enjoying a new life with the implanting of electrodes and the use of deep brain stimulation in other parts of the brain.

in in the hospital in a few weeks, but in the interim, having very little money, I arranged to stay at a YMCA; there was a large residential Y, I heard, in the Embarcadero, opposite the Ferry Building. The Y looked shabby, a little dilapidated, but comfortable and friendly, and I moved into a little room on the sixth floor.

Around 11:00 p.m., there was a soft knock on my door. I said, "Come in"; the door was not locked. A young man put his head round the door and, seeing me, exclaimed, "Sorry, I've got the wrong room."

"Don't be too sure," I answered, hardly believing what I was saying. "Why don't you come in?" He looked uncertain for a moment and then came in, locking the door behind him. This was my introduction to life at the Y—a continual opening and shutting of doors. Some of my neighbors, I observed, might have five visitors in a night. I had a peculiar, unprecedented feeling of freedom: I was no longer in London, no longer in Europe; this was the New World, and—within limits—I could do what I wished.

A few days later, Mount Zion said it had a room available for me, and I moved into the hospital—none the worse for my fling at the Y. *Interesting for me*

I spent the next eight months working for Levin and Feinstein; my official internship at Mount Zion would not begin until the following July.

Levin and Feinstein were as different from each other as could be imagined—Grant Levin unhurried and judicious, Feinstein passionate and intense—but they made a fine complementary pair, like my neurological chiefs Kremer and Gilliatt in London (and my surgical chiefs Debenham and Brooks at the Queen Elizabeth Hospital in Birmingham).

I had been fascinated by such partnerships even as a boy; in my chemical days, I read about the partnership of Kirchhoff and Bunsen and how their very different minds, together, were indispensable for the discovery of spectroscopy. I had been fascinated, at Oxford, by reading the famous paper on DNA by James Watson and Francis Crick and learning how different the two men were. And while I was toiling away at my uninspired internship at Mount Zion, I was to read about another seemingly improbable and incongruous pair of researchers, David Hubel and Torsten Wiesel, who were opening up the physiology of vision in the wildest, most beautiful way.

Besides Levin and Feinstein and their assistants and nurses, the unit employed an engineer and a physicist—there were ten of us in all—and the physiologist Benjamin Libet often visited.[2]

One patient in particular stayed in my mind, and I wrote to my parents about him in November of 1960:

> Do you remember a Somerset Maugham story about a man who had a spell cast on him by some jilted Island girl, and developed a fatal hiccup? One of our patients, a postencephalitic coffee baron, has had a hiccup for six days following operation, intractable to all the usual and some very unusual measures, and I fear may go the same way unless we block his phrenic nerves

---

2. It was at Mount Zion that Libet performed his astounding experiments showing that if subjects were asked to make a fist or perform another voluntary action, their brains would register a "decision" nearly half a second before there was any conscious decision to act. While his subjects felt that they had consciously and of their own free will made a movement, their brains had made a decision, seemingly, long before *they* did.

or something. I suggested bringing in a good hypnotist: I wonder
if this will work? Have you any experience of this as a major
problem?

My suggestion was viewed skeptically (I had doubts about it
myself), but Levin and Feinstein agreed to call in a hypnothera-
pist; nothing else had worked. To our amazement, he was able
to get the patient "under" and then to give him a posthypnotic
command: "When I snap my fingers, you will wake up and no
longer have hiccups."

The patient woke up, free from hiccups, and they never
recurred.

•

Though I had kept a journal in Canada, I stopped once I had
arrived in San Francisco and did not resume till I was on the
road again. I continued, however, to write long, detailed letters
to my parents, and in February of 1961 I wrote of seeing two of
my idols, Aldous Huxley and Arthur Koestler, at a conference
at UCSF:

Aldous Huxley gave a tremendous after dinner speech on Education. I had never seen him before, and was amazed by his
height and cadaverous emaciation. He is almost blind now, and
blinks his pebble like eyes constantly, as well as screwing up
his fists in front of them (this puzzled me, but I think now it
was to secure some pinhole vision): he has long corpse like hair
floating back, and a dun skin very loosely, very inaccurately,
covering the bony contours of his face. Leaning forward, in
intense concentration, he somewhat resembled Vesalius' skel-
eton in meditation. However, his marvelous mind is as good as
ever, served by a wit, a warmth, a memory and an eloquence

that brought everyone to their feet more than once. . . . And finally Arthur Koestler, on the creative process, a wonderful analysis so poorly audible and presented that half his audience walked out. Koestler, by the way, looks a little like Kaiser, a little like all the Hebrew teachers in the world, and speaks just like them [Kaiser, our Hebrew teacher, had been a familiar figure in the house from my earliest days]. Americans don't get lined, whereas Koestler's litvuk face was ostentatiously lined, seamed with great furrows of anguish and intelligence which seemed almost indecent in that smooth faced assembly!

Grant Levin, my genial and generous boss, had procured tickets for all of us in the neurosurgical unit to attend a conference called "The Control of Mind," and he frequently distributed tickets to musical, theatrical, and other cultural events in San Francisco, a rich diet which made me love the city more and more. I wrote to my parents of seeing Pierre Monteux with the San Francisco Symphony Orchestra:

He was conducting (always just one beat behind the orchestra, I felt), and the programme included Berlioz' *Symphonie Fantastique* (the execution scene always reminds me of that grisly Poulenc opera); "Till Eulenspiegel"; Debussy's "Les Jeux" (tremendous, could have been written by the early Stravinsky) and some minor nonsense by Cherubini. Monteux himself is almost ninety now, and has a wonderful pearlike shape, bobbing and waddling along, and a melancholy French moustache, somewhat like Einstein's. The audience went mad over him, partly I think out of appeasement (they hissed at him sixty years ago), and partly their faddy rather patronising mythomania, by which great age alone becomes a recommendation.

However, I admit it is exciting to think of the uncomputable rehearsals and first nights, the crushing failures, the heavenly successes, the tumbling billions of notes which must have run through that old brain in his ninety years.

I mentioned in the same letter that I had a strange experience when I went down to a Beat festival in Monterey:

My introduction to my host was strange: they said, "he's here," and led me into the bathroom. There I saw a sort of Christlike figure with his beard upraised in agony, clasping his arse under the hot shower. No doubt, the apparition of myself, black and gleaming, fresh from my bike, was equally novel and alarming to him. He had a painful perianal abscess, which I lanced for him with a brutish sailcloth needle sterilized by a match. There was a great gush of pus, and a loud bellow, followed by silence: he had fainted. When he came round he was much better, and I tasted the novel joy of being the practical man, the skillful physician-surgeon, who had helped the suffering artist. Later that day there was a crazy beatnik type party, at which young bespectacled women got up and recited poems about their bodies.

In England, one was classified (working class, middle class, upper class, whatever) as soon as one opened one's mouth; one did not mix, one was not at ease, with people of a different class—a system which, if implicit, was nonetheless as rigid, as uncrossable, as the caste system in India. America, I imagined, was a classless society, a place where everyone, irrespective of birth, color, religion, education, or profession, could meet each other as fellow humans, brother animals, a place where a professor could talk to a truck driver, without the categories coming between them.

I had had a taste, a glimpse, of such a democracy, an equality, when I roved about England on my motorcycle in the 1950s. Motorcycles seemed, even in stiff England, to bypass the barriers, to open a sort of social ease and good nature in everyone. "That's a nice bike," someone would say, and the conversation would go from there. Motorcyclists were a friendly lot; we waved to one another when we passed on the road, made conversation easily if we met at a café. We formed a sort of romantic classless society within society at large.

Finding that it made no sense to ship my motorbike from England, I decided to get a new one—a Norton Atlas, a scrambler which I could take off the road, onto desert tracks or mountain trails. I could keep it in the hospital courtyard.

I fell in with a group of fellow motorcyclists, and every Sunday morning we would meet in the city, go over the Golden Gate Bridge onto the narrow, eucalyptus-smelling road which wound up Mount Tamalpais, then along the high mountain ridge with the Pacific to our left, descending in wide swoops to have brunch together on Stinson Beach (or occasionally Bodega Bay, soon to be made famous by Hitchcock's film *The Birds*). Those early morning rides were about feeling intensely alive, feeling the air on one's face, the wind on one's body, in a way only given to motorcycle riders. Those mornings have an almost intolerable sweetness in memory, and nostalgic images of them are instantly provoked by the smell of eucalyptus.

On weekdays, I usually biked alone around San Francisco. But on one occasion, I approached a group—very different from our sedate and respectable Stinson Beach group—a noisy, uninhibited group, sitting on their bikes drinking cans of beer and smoking. When I got closer, I saw the Hells Angels logos on their jackets, but it was too late to turn around, so I drew up next to them and said, "Hello." My audacity and English accent intrigued them, as did, when they learned of it, my being a doctor. I was approved on the spot, without having to go through any rites of passage. I was pleasant, unjudgmental, and a doctor—and as such was called on, occasionally, to advise when riders were injured. I did not join them in any of their rides or other activities, and our mild, unexpected relation—unexpected for me, as for them—quietly petered out when I left San Francisco a year later.

•

If the twelve months between my leaving England and my starting my formal internship at Mount Zion had been full of adventure, unexpectedness, and excitement, being an intern there—rotating through a few weeks in medicine, in surgery, in pediatrics, etc.—was humdrum and boring by comparison, frustrating, too, because I had already done all this in England. I could not see further internship as anything but a bureaucratic waste of time, but all foreign medical graduates had to do two years of internship, irrespective of their previous training.

But there were advantages: I could stay another year in my beloved San Francisco at no cost; board and lodging were provided by the hospital. My fellow interns, from all over the States, were a varied and often gifted lot—Mount Zion had a high reputation, and this (combined with the opportunity of spending a year in San Francisco) was a strong attraction for newly qualified doctors—hundreds applied for internships at Mount Zion, and the hospital could afford to be highly selective.

I was especially close to Carol Burnett, a gifted black woman, a New Yorker who was fluent in many languages. On one occasion, we were both co-opted to scrub for a complex abdominal operation, though all we did was hold retractors and hand instruments to surgeons. There was no attempt to show us or teach us anything, and apart from snapping at us occasionally ("Forceps, quick!" "Hold the retractor tight!"), the surgeons ignored us. They talked a lot to each other, and at one point they dropped into Yiddish and made some ugly, slurring remarks about having a black intern in the operating room. Carol pricked up her ears at this and answered them in fluent Yiddish. The surgeons both turned red, and the operation came to a sudden halt.

"You never heard a *schwartze* speak Yiddish?" Carol added—a little extra, merry dig. I thought the surgeons were going to drop their instruments. Embarrassed, they apologized to Carol and were at pains to treat her with special consideration for the rest of our surgical rotation together. (We wondered whether the episode—and their getting to know and respect Carol as a person—would have a lasting effect on them.)

•

Most weekends, if I was not on duty, I would take off with my bike to explore Northern California. I was fascinated by the early gold-mining history of California; I had a particular feeling for Highway 49 and a tiny ghost town called Copperopolis, which I would pass on my way to the Mother Lode.

Sometimes I rode up the coast road, Highway 1, past the northernmost redwoods to Eureka and then on to Crater Lake in Oregon (I thought nothing then of riding seven hundred miles at a stretch). It was in the same year, otherwise monotonous with internship, that I discovered the wonders of Yosemite and Death Valley and made a first visit to Las Vegas, which one could see, in those unpolluted days, from fifty miles away, like a glittering mirage in the desert.

But if I made new friends in San Francisco, enjoyed the city, and toured widely on weekends, my neurological training was on hold, or would have been so had it not been for Levin and Feinstein, who invited me to conferences and allowed me to continue seeing their patients.

It was in 1958, I think, that my old friend Jonathan Miller gave me a book of poems by Thom Gunn—*The Sense of Movement*, which had just been published—and said, "You must meet Thom; he's your sort of person." I devoured it and determined that if I actually made it to California, the first thing I would do would be to look up Thom Gunn.

When I arrived in San Francisco, I made enquiries about Thom and was told that he was in England, on a fellowship at Cambridge. But a few months later, he had returned, and I met him at a party. I was twenty-seven, he was thirty or so; not such a great difference, but I was intensely conscious of his maturity and assurance, how he knew who he was, what his gifts were, what he was doing. He had by then published two books; I had never published anything. I thought of Thom as a teacher and mentor (though hardly a model, since our modes of writing were so different). I felt unformed, like a fetus, in comparison to him. In my nervousness, I mentioned to him that though I admired his poetry hugely, I had been disturbed by one of his poems called "The Beaters," with its sadomasochistic subject matter. He seemed embarrassed and reprimanded me delicately: "You mustn't confuse the poem with the poet."[3]

Somehow—I can no longer reconstruct exactly how—a friendship began, and a few weeks later I set out to call on him. Thom lived, in those days, at 975 Filbert, and that street, as San Franciscans know (but I did not), suddenly drops precipitously at a thirty-degree angle. I had my Norton scrambler and, rushing along Filbert, taking it far too fast, I suddenly found myself airborne, as in a ski jump. Fortunately, my bike took the jump easily, but I was rattled; it could have ended badly. When I rang Thom's bell, my heart was still pounding.

He invited me in, gave me a beer, and asked why I had been so eager to meet him. I said, simply, that many of his poems seemed to call to something deep inside me. Thom looked noncommittal. Which poems? he asked. Why? The first poem of his I had read was "On the Move," and as a motorcyclist myself, I said, I instantly resonated to it, as I had years before to T. E. Lawrence's short, lyrical piece "The Road." And I liked his poem titled "The Unsettled Motorcyclist's Vision of His Death," because I was convinced that, like Lawrence, I too would be killed on my motorbike.

I am not quite sure what Thom saw in me at this point, but I found in him great personal warmth and geniality mixed with fierce intellectual integrity. Thom, even then, was lapidary and incisive; I was centrifugal and effusive. He was incapable of indirection or deceit, but his directness was always accompanied, I thought, by a sort of tenderness, too.

---

3. I was interested to see that when Thom's *Collected Poems* came out in 1994, this was the one poem from *The Sense of Movement* that he chose not to reprint.

Thom sometimes gave me manuscripts of new poems. I loved the contained energy in them—the holding in, the binding, of unruly energy and passion by the strictest, the most controlled of poetic forms. My favorite among his new poems, perhaps, was "The Allegory of the Wolf Boy" ("At tennis and at tea/Upon the gentle lawn, he is not ours,/But plays us in a sad duplicity"). This corresponded to a certain duplicity I felt in myself, which I thought of in part as a need to have different selves for day and night. By day I would be the genial, white-coated Dr. Oliver Sacks, but at nightfall I would exchange my white coat for my motorbike leathers and, anonymous, wolf-like, slip out of the hospital to rove the streets or mount the sinuous curves of Mount Tamalpais and then race along the moonlit road to Stinson Beach or Bodega Bay. This doubleness was assisted by my having the middle name of Wolf; for Thom and my bike friends, my name was Wolf, where for my fellow doctors it was Oliver. In October of 1961, Thom gave me a copy of his new book, *My Sad Captains,* and inscribed it "to the Wolf Boy (no allegorizing needed!), with *alles gute,* and admiration, from Thom."

•

In February of 1961, I wrote to my parents that I had got my green card and was now a bona fide immigrant—a "resident alien"—and had declared my intention of becoming a citizen, which one could do without forfeiting British citizenship.[4]

I mentioned too that I would shortly be taking the state boards—a fairly comprehensive exam for foreign medical grad-

---

4. The intention was genuine, but more than fifty years have passed, and I am still not a citizen. It was similar with my brother in Australia. He arrived there in 1950 but only got Australian citizenship fifty years later.

uates, to see if they were really up to scratch on their basic science as well as their medicine.

I had written to my parents, back in January, that I was contemplating "an immense journey across the States, and back via Canada, even detouring to Alaska, between my exams and starting internship: probably around 9000 miles in all. It should be a unique opportunity to see the country, and visit other universities."

And now, with the state boards passed and a more suitable motorbike—I traded in my Norton Atlas for a secondhand BMW R69—I was ready to set out. My time off had been whittled down, and I could no longer include Alaska in my round-America circuit. I wrote again to my parents:

I have drawn a great red line on my map: Las Vegas, Death Valley, Grand Canyon, Albuquerque, Carlsbad Caverns, New Orleans, Birmingham, Atlanta, Blue Ridge Parkway to Washington, Philadelphia, New York, Boston. Up through New England to Montreal, sidetrack to Quebec. Toronto, Niagara Falls, Buffalo, Chicago, Milwaukee. The Twin Cities, then up to Glacier and Waterton National Parks, down to Yellowstone Park, Bear Lake, Salt Lake City. Back to San Francisco. 8000 miles. 50 days. $400. If I avoid: sunstroke, frostbite, imprisonment, earthquakes, food poisoning and mechanical disaster— why, it should be the greatest time of my life! Next letter from the road.

When I told Thom of my travel plans, he suggested that I keep a journal—a portrait of my experiences, "Encountering America"—and that I send it to him. I was on the road for two months and filled several notebooks, posting them to Thom one by one. He seemed to like my descriptions of people and places, sketches and scenes, and he thought that I had a gift for observation, though he sometimes took me to task for my "sarcasms and grotesqueries."

One of the journals I sent to him was "Travel Happy."

TRAVEL HAPPY (1961)

A few miles north of New Orleans, my bike gave out. I pulled over and began tinkering with the engine in some forsaken lay-by. As I lay there on my back I detected, with some sixth seismic sense, a distant tremor, like a far-off earthquake. It advanced towards me, becoming a rattle, then a rumble, and finally a roar, culminating in the screech of air brakes and a terrific cheerful honking. I looked up, paralyzed, and saw the vastest truck I had ever seen, a very Leviathan of the road. An impudent Jonah stuck his head out from the window and hollered at me from the great altitude of his cab.

"Anything I can do?"

"She's shot!" I answered. "Busted rod or something."

"Shit!" he remarked pleasantly. "If that breaks loose, it'll cut your leg off! Be seeing you."

He grimaced, ambiguously, and maneuvered his huge truck onto the road once more.

I rode on and on, and soon left the swampy lowlands of the Delta. Soon I was in Mississippi. The road meandered here and there, capricious and unhurried, winding through thick forests and open pastures, through orchards and meadows, over half a dozen intersecting rivers, and in and out of farms and villages, all tranquil and motionless in the morning sun.

But after I crossed into Alabama the bike grew rapidly worse. I hung on every variation of its sound, pondering on noises which were sinister but unintelligible. It was disintegrating fast, this much was certain; but, ignorant and fatalistic, I felt I could do nothing to arrest its fate.

Five miles beyond Tuscaloosa the engine faltered and
seized. I grabbed the clutch, but one of the cylinders was
already smoking by my foot. I dismounted and laid the bike
out flat upon the ground. Then I advanced towards the roadside, holding a clean white handkerchief in my left hand.

The sun was dropping in the heavens, and an icy wind
sprang up. The traffic was diminishing.

I had almost given up hope and was waving quite mechani-
cally, when abruptly, incredulously, I realized that a truck was
stopping. It looked familiar. Narrowing my eyes I spelled out
its registration: 26539, Miami, FLA. Yes, that was it: the vast
truck which had stopped for me this morning.

As I ran up to it, the driver descended from his cab, nodded
towards the bike and grinned:

"So you finally fucked it up, huh?"

A boy followed him down from the truck, and together we
scrutinized the wreck.

"Any chance of a tow to Birmingham?"

"Naah, law says no!" He scratched the stubble on his chin,
then winked: "Let's heave the motorcycle inside!"

We struggled and panted as we hoisted the heavy machine
into the belly of the truck. Finally, it was secured among the
furniture, tethered with ropes, and hidden from prying eyes by
a tumbled mass of sacking.

He climbed back into the cab, followed by the boy, and then
myself, and we ensconced ourselves—in this order—along
its broad seat. He gave a little bow, and performed formal
introductions:

"This is my trucking partner, Howard. What's your name?"

"Wolf."

"Mind if I call you Wolfie?"

"No, you go right ahead. And yours?"

"Mac. We're all Mac, you know, but I'm the genuine original Mac! You can see it on my arm."

For a few minutes we drove in silence, taking stock of one another, surreptitiously.

Mac looked about thirty, though he could have been five years either side of this. He had a vigorous, alert, and handsome face, with a straight nose, firm lips, and a clipped moustache. He could have been a British cavalry officer; he could have played small romantic parts on screen or stage. These were my first impressions.

He wore the peaked and crested cap all truckers wear, and a shirt emblazoned with the name of his company: ACE TRUCKERS, INC. Upon his arm was a red badge with the legend: PLEDGED TO COURTESY AND SAFETY, and half-concealed under the rolled-up sleeve, his name: MAC, entwined with a struggling python.

Howard could have passed for sixteen, but for the set lines which arched above his mouth. His lips were always slightly parted, revealing large yellow teeth, irregular but powerful, and an astonishing expanse of gum. His eyes were of the palest blue, like the eyes of some albino animal. He was tall and well-built, but graceless.

After a while he turned his head and gazed at me with his pale animal eyes. First he stared straight into my own eyes for a minute; then his gaze widened to embrace the rest of my face, my visible body, the cab of the truck, and the monotonously moving road outside the window. As his attention widened, so it faded, until his face resumed once more its vacant dreaminess. The effect was first disquieting and then uncanny. With a sudden horror and pity, I realized that Howard was feeble-minded.

Mac gave a short laugh in the darkness. "Well, think we make a good pair?"

"I'll soon see," I answered. "How far will you be taking me?"

"To the ends of the earth, New York anyhow. We'll make it Tuesday, maybe Wednesday."

He lapsed into silence again.

A few miles later he asked me suddenly, "Ever hear of the Bessemer process?"

"Yes," I said. "We 'did' it in chemistry at school."

"Ever hear of John Henry, the steel-driving nigger? Well, he lived right here. When they made a machine to drive a steel pick into a riverbed, they said that human labor could never compete. The niggers made a wager, and brought up their strongest man: John Henry. They say his arms was bigger 'n twenty inches. He held a mallet in each hand, and he drove in a hundred picks quicker than their machine. Then he lay down and died. Yessir! This is steel country."

We were surrounded by scrap yards, auto wreckers, railway sidings, and smelting works. The air resounded with the clangor of steel, as if the whole of Bessemer was some gigantic forge or armory. Torches of flame topped the high chimneys, roaring as they swept up from the furnaces below.

I had only once before seen a city illuminated by flames, and that was as a child of seven, when I saw London in the Blitz of 1940.

After we had negotiated Bessemer and Birmingham, Mac started to speak freely of himself.

He had bought his truck for $500 down, and the balance—$20,000—over a year. He could take up to 30,000 pounds of cargo, and traveled anywhere and everywhere: Canada, the States, Mexico, so long as there were decent roads and money to be made. He averaged four hundred miles in a ten-hour working day; it was illegal to work longer at a stretch, though frequently done. He had trucked, on and off, for twelve years now, and had been "riding double" with Howard for just six months. He was thirty-two and lived in Florida, he had a wife and two children, and he made $35,000 a year, he said.

He ran away from school when he was twelve years old, and, looking older, got himself a traveling salesman's job. At seventeen, he joined the police force, and at twenty was a considerable firearms expert. That year he'd been involved in a gun fight and narrowly escaped being shot in the face at point-blank range. He lost his nerve after that and changed to trucking, though he was still an honorary member of the Florida police force, and received one dollar a year in token of this.

Had I ever been in a gun fight? he enquired. No. Well, he'd been in more than he could remember, both as a police-man and a trucker. I'd find his "trucker's friend" right under the seat, if I cared to look; they all carried guns on the road. Though the best weapon in an unarmed fight was a piece of piano wire. Once you had it round your opponent's neck there was nothing he could do. You gave a little tug—and the head fell off: easy—like cutting cheese! There was no mistaking the relish in his voice.

He had carried everything in trucks, from dynamite to prickly pears, but had now settled down to trucking furniture, although this included anything which a man might keep in his house. He had the contents of seventeen homes on board, including seven hundred pounds of weights (the property of a muscleman moving out of Florida); a grand piano made in Germany, said to be the best there was; ten television sets (they had one out at the truck stop last night and plugged it in); and an antique four-poster on its way to Philadelphia. If I wished to, I might sleep in this at any time.

The four-poster brought a nostalgic smile to his face, and he started talking of his sexual exploits. He seemed to have had an incredible success at all times and places, though four women held pre-eminence in his affections: a girl in L.A. who once eloped with him as a stowaway in this truck; two maiden ladies in Virginia he slept double with, who had showered him for years with clothes and money; and a nymphomaniac in Mexico City, who could take twenty men in a night and still cry for more.

As he warmed up, the last traces of diffidence disappeared, and he emerged as the full-blown Sexual Athlete and Storyteller. He was God's gift to lonely women.

It was during this recital that Howard, who had been lying in a sort of stupor, pricked up his ears and showed his first signs of animation. Mac, seeing this, first humored him, then started to goad him with a teasing banter: tonight, he said, he was going to get a girl in the cab and lock Howard in the trailer, though one night, if the boy looked sharp, he might procure a real whore (he pronounced it "hooorrh") for him. Howard grew hot and wild, and started panting with excitement; finally he lunged angrily at Mac.

As they struggled in the cab, half in fury, half in play, the steering wheel was jolted violently, and the huge truck rocked dangerously along the road.

But between his taunts Mac was also educating Howard, informally:

"What's the capital of Alabama, Howard?"

"Montgomery, you filthy sonofabitch!"

"Yeah, that's right. 'Tain't always the biggest cities that's the state capital. And those are pecan trees, look—over there!"

"Fuck you, I don't care!" grumbled Howard, but craned his head to see them, nonetheless.

An hour later we pulled in at a truck stop somewhere in the wilds of Alabama, for Mac had decided we should stay here overnight: it was called "Travel Happy."

We went inside for coffee. Mac settled down, with polite determination, to entertain me with "funny stories," of which he had an endless, execrable store, much inferior to his own first-hand experiences. Having discharged this friendly duty, he wandered off to join the crowd around the jukebox.

Truckers always gather round a jukebox on Saturday evenings, and try desperately to make it to the truck stop on this one night. The jukebox at Travel Happy, in particular, enjoys a certain fame, for it has a magnificent collection of trucking songs and ballads, epics of the road: savage, bawdy, melancholy, or wistful, but all with an insistent energy and rhythm, a special excitement which spells the very poetry of motion and endless roads.

Truckers are generally solitary men. Yet occasionally—as in a hot and crowded truckers' café, listening to some infinitely familiar record blaring on the jukebox—they are stirred, transfigured suddenly, without words or actions, from an inert

crowd to a proud community: each man still anonymous and
transient, yet knowing his identity with those around him,
all those who came before him, and all who are figured in the
songs and ballads.

Tonight Mac and Howard had become, like all the others,
rapt and proud, in unwitting transcendence of themselves.
They were sinking into a timeless reverie.

Around midnight Mac gave a violent start and then tugged
Howard by the collar. "Okeydoke, kiddo," he said, "let's find
ourselves a place to sleep. Wanna say the trucker's prayer
before you turn in?"

He took a creased card from his pocket-book, and handed it
to me. I flattened it out and read aloud:

> *Oh Lord give me strength to make this run*
> *For U.S. currency and not for fun.*
> *Please help me not to have a flat*
> *No engine trouble or likes of that.*
> *Help me pass the scale and the ICC*
> *Or make the JP make me go free.*
> *Keep the Sunday drivers out of my way*
> *And the woman drivers too, I pray.*
> *And when I wake in the stinking cab*
> *Let me wake where there's ham and eggs to grab.*
> *Make the coffee strong and the women weak*
> *And the waitress cute, and not some freak.*
> *Make the highway better and the gasoline cheaper*
> *And on my return, Lord, get me a sleeper.*
> *If you'll do this, Lord, with a little luck*
> *I'll keep right on driving a darned old truck.*

Mac took his blanket and pillow into the cab, Howard crept into a nook among the furniture, and I bedded down in a heap of sacks beside the bike (the promised four-poster was at the front, inaccessible).

I closed my eyes and sharpened my ears. Mac and Howard were whispering to one another, using the truck's solid walls as a conductor. Putting my ear to part of the latticed framework I now heard other noises too—the sounds of joking and drinking and making love—coming from all the other trucks all around us, impinging on the antenna of my ear.

I lay, contented, in the darkness, feeling myself in a very aquarium of sound; and very soon I feel asleep.

Sunday was a day of rest, at Travel Happy as all over America. A pane of lighted glass above me, smell of straw and sacking, smell of the leather jacket which serves me as a pillow. For a moment of confusion I fancied myself in a great barn somewhere, and then I instantly remembered.

I heard a gentle sound of running water, which started suddenly and ended gradually, lingeringly, with two small afterspurts. Someone was pissing against the side of the truck; of *our* truck, I thought, with a new possessiveness. I scrambled out from under the sacks and tiptoed to the door. A steaming trail from wheel to ground bore testament to the crime, but the offender had crept away.

It was seven o'clock in the morning. I seated myself on the high step of the cab and started to scribble in my journal. A shadow fell across the page; I looked up and recognized a trucker seen dimly in the smoke-filled café the night before. It was John, the blond Lothario from the "Mayflower Transit Co.," perhaps the very man who had pissed against our wheel. We chatted for a while, and he told me he had left Indianapolis—our immediate destination—the night before: it had been snowing there.

A few minutes later, another trucker shambled up, a short fat man wearing the floral shirt of the "Tropicana Orange Juice Co. Fla.," half-unbuttoned to expose a hairy pudding belly.

"Christ, it's cold here," he muttered. "It was ninety in Miami yesterday!"

Others gathered round me, talking of their routes and journeys, of mountains, oceans, plains; of forest and desert; of snow and hail and thunder and cyclone—all encountered within the span of a single day. A world of traveling and strange experience was gathered at Travel Happy on this, and every, night.

I walked round to the back of the truck and saw through its half-open doors Howard asleep in his niche. His mouth hung open, and his eyes too—I noticed with a qualm—were not fully closed. I thought for a moment he had died in the night, until I saw him breathe and twist a little in his sleep.

An hour later Mac awoke, tousled and disheveled, and staggered from the cab; he vanished in the direction of the truck-stop "bunkhouse," carrying with him a massive Gladstone bag. When he came back a few minutes later, he was perfectly groomed and shaven, coolly and cleanly dressed for the Lord's day.

I joined him and we walked towards the cafeteria.

"What about Howard?" I asked. "Shall I wake him now?"

"No. The kid'll wake up later."

Mac obviously felt the need to talk to me without his being around.

"He'd sleep all day if I let him," he grumbled over breakfast. "He's a good kid, you know, but not too bright."

He had met Howard six months before—a bum of twenty-three—and taken pity on him. The boy had run away from home ten years before, and his father—a well-known banker in Detroit—made little effort to retrieve him. He had taken to the roads and wandered round at large, doing a little casual work from time to time; occasionally begging, occasionally stealing, and managing to avoid churches and prisons. He was briefly in the army, but soon discharged as feeble-minded.

Mac picked him up in the truck one day and "adopted" him: he took him along now on every trip, showing him the country, teaching him how to pack and crate (and also how to talk and act), and paying him a regular salary. When they returned to Florida, after a journey's end, Howard would stay with Mac's wife and family, where he had the status of a younger brother.

As we drank our second cup of coffee, Mac's handsome face grew clouded:

"He won't be with me much longer, I imagine. Maybe I won't even be driving myself for too much longer."

He explained that he had a curious "accident" some weeks ago, when he blacked out without warning and his truck had run into a field. The insurance paid, but insisted that he would have to have a medical examination; they further objected to his having a trucking partner, whatever the outcome of the examination.

Clearly Mac fears, as his insurance company must suspect,

that his "blackout" was due to epilepsy, and that the medical will see the end of his driving days. He has had the foresight to line up a good job, in insurance, in New Orleans.

Howard walked in at this point, and Mac rapidly changed the subject.

After breakfast Mac and Howard sat on a discarded tire, shying stones at a wooden post. We talked vaguely and incoherently of many things, spelling away the gentle Sunday indolence of a trucker's lot. After a couple of hours they got bored, and climbed back into the truck to sleep again.

I grabbed a couple of burlaps from the trailer and settled down to sunbathe, surrounded by broken bottles, sausage skins, food, beer cans, decaying contraceptives and an incredible litter of torn and screwed-up paper: here and there a stalk of wild onion or lucerne was poking through the rubble.

As I lay and dozed or wrote, my thoughts turned often to food. Behind me were a score of beggarly chickens scrabbling in the dust, which I gazed at from time to time with a wistful sigh, for Mac had waved his "trucker's friend" (an efficient-looking automatic) at them earlier, saying:

"Poultry for dinner tonight!" with a pleasant chuckle.

Every hour or so I would get up to stretch my legs, and consume four coffees and a black-walnut ice cream in the café, leading to my present total of twenty-eight and seven, respectively.

I have also paid many visits to the bunkhouse washroom, having had an incandescent diarrhea since tasting Mac's hot peppers last night.

There are five contraceptive machines in the little room, an interesting example of how commercial pressures will follow a man into his most intimate activities. The cost of these beautiful articles ("electronically rolled, cellophane-sealed, supple, sensitive, and transparent," as they were rapturously described) was three for half a dollar, though this had been modified, clumsily, to read: THREE FOR A DOLL. There was also a machine called Prolong, which dispensed an anesthetic ointment designed, it was stated, "to aid in the prevention of premature climax." But John, the blond Lothario, who is turning out to be a veritable compendium of sexual information, says that piles ointment is much better. Prolong is too strong—you never know whether you have come or not.

In the middle of the afternoon Mac suddenly announced that we would be staying at Travel Happy for another night. He wore a pleased, deliberately mysterious smile—no doubt he has lined up Sue or Nell for an assignation in the cab tonight. Howard has been behaving like an excited dog in this atmosphere of intrigue. Despite his brave show I suspect (and Mac has confirmed) that he has never yet made out with a girl. Indeed, Mac has procured him girls from time to time, but Howard—so vociferous in imaginary achievement—becomes timid and boorish when confronted with the reality, and things always "fall through" at the very last moment.

I returned to my writing and my cups of coffee. Occasionally, I walked outside to stretch my legs, and to peer curiously at all the truckers round me, snoring in their cabs, comparing their faces and their postures in repose.

At 4:20 the dawn appeared, dim and indecisive in the east. One trucker woke up and walked towards the bunkhouse to take a leak. Returning to his truck he checked over his cargo, pulled himself into the cab, and slammed the door. He started his engine with a roar, and slowly lumbered out. The other trucks remained silent and sleeping.

By five o'clock the stillborn dawn had been replaced by a fine and drizzling rain. One of the ragged cockerels was kicking up a din, and the twitter of insects had started in the grass.

Six o'clock, the café is filled with the smell of hotcakes and butter, bacon and eggs. The night waitresses take their leave, wishing me good luck in my travels around America. The day staff troop in, and smile to see me seated at the table I occupied all yesterday.

I can come and go now as I please in the little café. They no longer charge me for anything. I have drunk more than seventy cups of coffee in the past thirty hours, and this achievement deserves some small concession.

Eight o'clock, and Mac and Howard have just hurried off to downtown Coleman, to help the Mayflower men unpack their cargo. The pace is suddenly different, for today they have said nothing, they have skipped breakfast, and they haven't washed. Mac's Gladstone bag has been put away for another week.

I crawled into the cab which Mac had just vacated—it was still warm from his humid sleep—covered myself with his old worn blanket, and in a moment fell asleep myself. I was awoken briefly, at ten, by a heavy fusillade of rain upon the roof, but there was no sign yet of Mac or Howard.

They finally turned up at half past twelve, heavy-footed and bedraggled from shifting the heavy cargo in a rainstorm.

"Christ!" said Mac. "I'm shot. Let's eat—and we'll be on our way in an hour."

This was three hours ago, and still we have not moved! They have been smoking and boasting and fiddling and flirting, as if an unhurried thousand years lay before them. Mad with impatience, I withdrew to the cab with my notebooks. Lothario John tried to mollify me:

"Take it easy, kiddo! If Mac says he'll make New York by Wednesday, he'll do it, even if he stays at Travel Happy till Tuesday evening."

After forty hours here, this truck stop has become infinitely familiar. I know a score of men—their likes and dislikes, their jokes and idiosyncrasies. And they know mine, or think they do, and call me "Doc" or "Professor" indulgently.

I know all the trucks—their tonnage and cargoes, their performances and quirks, and their insignia.

I know all the waitresses at Travel Happy—Carol, the boss, has taken a Polaroid snap of me standing between Sue and Nell, my face unshaven and dazzled in the flash-light. She has stuck it up along with her other photos, so that now I have my place in her thousand-brothered family, her "boyfriends" who come and go on the long cross country trucking routes.

"Yeah!" she will say to some future, puzzled customer who scrutinizes the print. "That's 'Doc.' Great guy he was, bit strange maybe. He rode with Mac and Howard, those two over there. I often wonder what became of him."

Muscle Beach

When I finally made it to New York in June of 1961, I borrowed money from a cousin and bought a new bike, a
BMW R60—the trustiest of all the BMW models. I wanted no
more to do with used bikes, like the R69 which some idiot or
criminal had fitted with the wrong pistons, the pistons that
had seized up in Alabama.

I spent a few days in New York, and then the open road
beckoned me. I covered thousands of miles in my slow, erratic
return to California. The roads were wonderfully empty, and
going across South Dakota and Wyoming, I would scarcely see
another soul for hours. The silence of the bike, the effortless-
ness of riding, lent a magical, dreamlike quality to my motion.

There is a direct union of oneself with a motorcycle, for it
is so geared to one's proprioception, one's movements and postures, that it responds almost like part of one's own body. Bike
and rider become a single, indivisible entity; it is very much
like riding a horse. A car cannot become part of one in quite
the same way.

I arrived back in San Francisco at the end of June, just in
time to exchange my bike leathers for the white coat of an
intern in Mount Zion Hospital.

During my long road trip, with snatched meals here and
there, I had lost weight, but I had also worked out when possible at gyms, so I was in trim shape, under two hundred pounds,

when I showed off my new bike and my new body in New York in June. But when I returned to San Francisco, I decided to "bulk up" (as weight lifters say) and have a go at a weight-lifting record, one which I thought might be just within my reach. Putting on weight was particularly easy to do at Mount Zion, because its coffee shop offered double cheeseburgers and huge milk shakes, and these were free to residents and interns. Rationing myself to five double cheeseburgers and half a dozen milkshakes per evening and training hard, I bulked up swiftly, moving from the mid-heavy category (up to 198 pounds) to the heavy (up to 240 pounds) to the superheavy (no limit). I told my parents about this—as I told them almost everything—and they were a bit disturbed, which surprised me, because my father was no lightweight and weighed around 250 himself.[1]

I had done some weight lifting as a medical student in London in the 1950s. I belonged to a Jewish sports club, the Maccabi, and we would have power-lifting contests with other sports clubs, the three competition lifts being the curl, the bench press, and the squat, or deep knee bend.

Very different from these were the three Olympic lifts—the press, the snatch, and the clean and jerk—and here we had world-class lifters in our little gym. One of them, Ben Helf-gott, had captained the British weight-lifting team in the 1956 Olympic Games. He became a good friend (and even now, in his eighties, he is still extraordinarily strong and agile).[2] I tried

---

1. My father would eat continually in the presence of food but go all day without food if it was not available; it is similar with me. In the absence of internal controls, I have to have external ones. I have fixed routines for eating and dislike deviations from them.

2. Helfgott's achievement was all the more extraordinary because he had survived the camps at Buchenwald and Theresienstadt.

the Olympic lifts, but I was too clumsy. My snatches, in particular, were dangerous to those around me, and I was told in no uncertain terms to get off the Olympic lifting platform and go back to power lifting.

Besides the Maccabi, I trained occasionally at the Central YMCA in London, which had a weight-lifting gallery supervised by Ken McDonald, who had lifted for Australia in the Olympic Games. Ken carried a good deal of weight himself, especially in his lower half; he had enormous thighs and was a world-class squatter. I admired his squatting ability, and I too wanted to develop such thighs and to develop the back power crucial for squatting and overhead lifts. Ken favored stiff-legged deadlifting—a lift designed, if any lift is, to injure the back, for the entire load is focused on the lumbar spine and not taken, as it should be, by the legs. As I improved under his tutelage, Ken invited me to join him in a weight-lifting exhibition—the two of us alternating our dead lifts. Ken deadlifted 700 pounds. I could only manage 525, but this was applauded, and I felt a brief pleasure and pride that despite being a novice, I had been able to companion Ken on his record-breaking dead lift. My pleasure was very short-lived, for a few days later I developed a pain in my lower back so intense that I could hardly move or breathe; I wondered if I had fractured a vertebra. Nothing amiss was seen on X-ray, and the pain and spasms resolved in a couple of days, but I was to have excruciating back attacks for the next forty years (they only let up, for some reason, when I was sixty-five, or were then, perhaps, "replaced" by sciatica).

My admiration of Ken's training schedule extended to the special, largely liquid diet he had designed in order to bulk up. He would come to his workouts with a half-gallon bottle filled with a thick, treacly mixture of molasses and milk supplemented by assorted vitamins and yeast. I decided to do the same, but I overlooked the fact that yeast ferments sugar, given enough time. When I pulled my bottle out of my gym bag, it was bulging ominously. It was clear that the yeast had fermented the mixture; I had put it in hours beforehand, while Ken (as he told me later) sprinkled it in just before he came to the gym. The contents of the bottle were under pressure, and I felt a little frightened; it was as if I had suddenly found myself in possession of a bomb. I thought that if I unscrewed the top very slowly, there could be a gentle decompression, but as soon as I loosened the top a little, it blew; the whole half gallon of sticky black (and now slightly alcoholic) muck went up like a geyser high into the air, landing all over the gym. There was laughter at first and then rage, and I was sternly warned never again to bring anything but water to the gym.

•

The Central YMCA in San Francisco had particularly good weight-lifting facilities. The first time I went there, my eye was caught by a bench-press bar loaded with nearly 400 pounds. No one at the Maccabi could bench-press anything like this, and when I looked around, I saw no one in the Y who looked up to such a weight. No one, at least, until a short but hugely broad and thick-chested man, a white-haired gorilla, hobbled into the gym—he was slightly bowlegged—lay down on the bench, and, by way of warm-up, did a dozen easy reps with the bench-press bar. He added weights for subsequent sets, going to nearly 500 pounds. I had a Polaroid camera with me and took a picture as he rested between sets. I got talking to him later; he was very genial. He told me that his name was Karl Norberg, that he was Swedish, that he had worked all his life as a longshoreman, and that he was now seventy years old. His phenomenal strength had come to him naturally; his only exercise had been hefting boxes and barrels at the docks, often one on each shoulder, boxes and barrels which no "normal" person could even lift off the ground.

I felt inspired by Karl and determined to lift greater poundages myself, to work on the one lift I was already fairly good at—the squat. Training intensively, even obsessively, at a small gym in San Rafael, I worked up to doing five sets of five reps with 555 pounds every fifth day. The symmetry of this pleased me but caused amusement at the gym—"Sacks and his fives." I didn't realize how exceptional this was until another lifter encouraged me to have a go at the California squat record. I did so, diffidently, and to my delight was able to set a new record, a squat with a 600-pound bar on my shoulders. This was to serve as my introduction to the power-lifting world; a weight-lifting record is equivalent, in these circles, to publishing a scientific paper or a book in academia.

•

By the spring of 1962, my internship at Mount Zion was coming to an end, and my residency at UCLA was due to begin on July 1. But I needed time to visit London before the residency started. I had not seen my parents for two years, and my mother had just broken a hip, so I was very glad to be with her soon after her operation. Ma dealt with the trauma, the surgery, and the ensuing weeks of pain and rehabilitation with great fortitude and was determined to get back to seeing her own patients as soon as she was off crutches.

Our winding staircase, with worn carpet and sometimes loose stair rods, was not safe for someone on crutches, so I carried her up and down as she needed—she had been against my heavy lifting, but now she was glad of my strength—and I delayed my return until she could manage the stairs herself.[3]

At UCLA, we residents had a weekly "Journal Club"; we would read the latest papers in neurology and discuss them. I sometimes annoyed the group, I think, by saying that we should also discuss the writings of our nineteenth-century forebears, relating what we were seeing in patients to *their* observations and thoughts. This was seen by the others as archaism; we were short of time, and we had better things to do than consider such "obsolete" matters. This attitude was reflected, implicitly, in many of the journal articles we read; they made little reference to anything more than five years old. It was as if neurology *had* no history.

I found this dismaying, for I think in narrative and historical terms. As a chemistry-mad boy, I devoured books on the history of chemistry, the evolution of its ideas, and the lives of my favorite chemists. Chemistry had, for me, a historical and human dimension too.

It was similar when my interests moved from chemistry to biology. Here, of course, my central passion was for Darwin, and I read not only the *Origin* and the *Descent*, not only *The Voyage of the* Beagle, but all his botanical books, too, with *Coral Reefs* and *Earthworms* thrown in. I loved his autobiography most of all.

----

3. Unfortunately, her hip had broken in a bad place, compromising the blood supply in the head of the femur. This caused it to undergo so-called avascular necrosis and finally collapse, causing intense, unremitting pain. Though my mother was stoical and continued seeing patients, leading a full life despite the pain, it aged her, and when I returned to London again in 1965, she looked ten years older than she had three years earlier.

Eric Korn had a similar passion and eventually gave up a career in academic zoology to become an antiquarian book-seller specializing in Darwiniana and nineteenth-century science. (He was consulted by booksellers and scholars the world over for his unique knowledge of Darwin and his times, and he was a good friend of Stephen Jay Gould's. Eric was even asked—no one else could have done it—to reconstruct Darwin's own library in Down House.)

I was not a book collector myself, and when I bought books or articles, it was to read them, not to show them. So Eric reserved his torn or damaged books for me, books which might lack a cover or a title page—books no collector would want but which I could afford to buy. As my interests moved to neurology, it was Eric who got me Gowers's 1888 *Manual*, Charcot's *Lectures*, and a host of lesser-known but to me beautiful and inspiring nineteenth-century texts. Many of these became central to the books I was later to write.

•

I was fascinated by one of the first patients I saw at UCLA. It is not uncommon to have a sudden myoclonic jerk as one is falling asleep, but this young woman had a much more severe myoclonus and would react to flickering light of a certain frequency with sudden convulsive jerks of her body or, occasionally, with full-blown seizures. These problems had been in her family for five generations. With my colleagues Chris Herrmann and Mary Jane Aguilar, I wrote my first paper (for the journal *Neurology*) about her, and, fascinated by myoclonic jerks and the many conditions and circumstances in which these could occur, I wrote a small book about them all, which I titled "Myoclonus."

When, in 1963, Charles Luttrell, a neurologist noted for his outstanding work on myoclonus, visited UCLA, I told him of my own interest in the subject and said I would be most grateful for his opinion on my little book. He was agreeable, and I gave him the manuscript; I did not have a copy. A week passed, and another, and another, and at six weeks I could contain my impatience no longer and wrote to Dr. Luttrell. I learned that he had died. I was shocked. I wrote a letter of condolence to Mrs. Luttrell, in which I spoke of my admiration for her husband's work, but I thought it would be unseemly, at this juncture, to ask for the return of my manuscript. I never asked, and the manuscript was never returned. I do not know whether it is still in existence; it was probably tossed out, but perhaps it still resides quietly in some forgotten drawer.

•

In 1964, I saw a puzzling young man, Frank C., in the neurology clinic at UCLA. He had incessant jerking movements of his head and limbs which had started when he was nineteen, getting gradually worse over the years; his sleep, more recently, had been disturbed by massive jerks of his entire body. He had tried tranquilizers, but nothing seemed to help the jerks, and Frank, depressed, had started drinking heavily. His father, he said, had had identical movements starting in his early twenties, had become depressed and alcoholic, and had finally committed suicide at the age of thirty-seven. Frank himself was thirty-seven now, and he said he could imagine exactly how his father felt; he feared that he might take the same step himself.

He had been in the hospital six months earlier, where various diagnoses were considered—Huntington's chorea, postencephalitic parkinsonism, Wilson's disease, etc.—but none of these could be confirmed. So Frank and his strange disease presented an enigma. I gazed at his head at one point, thinking, "What's going on inside there? I wish I could *see* your brain."

Half an hour after Frank had left the clinic, a nurse rushed in and said, "Dr. Sacks, your patient has just been killed—hit by a truck—he died instantly." An immediate autopsy was performed, and two hours later I had Frank's brain in my hands. I felt awful—and guilty. Could my wish to see his brain have played a part in his fatal accident? I could not help wondering, too, whether he had decided to end things and stepped deliberately in front of the truck.

His brain was of normal size and showed no gross abnormalities, but when I got some slides under the microscope a few days later, I was startled to see gross swelling and tortuosity of nerve axons, pale spherical masses, and rusty brown pigmentation from iron deposits in the substantia nigra, the pallidum, and the subthalamic nuclei—all parts of the brain that regulate movement—and nowhere else.

I had never before seen such huge swellings confined to the nerve axons or bits of detached axon; these did not occur in Huntington's disease or any other disease I had encountered. But I had seen pictures of such axonal swellings, photographs illustrating a very rare disease described by two German pathologists, Hallervorden and Spatz, in 1922—a disease that started in youth with abnormal movements but then, as it advanced, caused widespread neurological symptoms, dementia, and finally death. Hallervorden and Spatz observed this fatal disease in five sisters. At autopsy, their brains contained axonal swellings and lumps of detached axon, as well as brownish discolorations in the pallidum and substantia nigra.

It looked, then, as if Frank might have had Hallervorden-Spatz disease and that his tragic death had allowed us to see the neural basis of this at a very early stage.

If I was right, we had a case which exemplified, better than any previously described, the initial and fundamental changes in Hallervorden-Spatz disease, uncontaminated by all the secondary features seen in more advanced cases. I was intrigued by the strangeness of a pathology which seemed aimed only at the nerve axons, leaving the bodies of the nerve cells and the myelin sheaths of the axons untouched.

Just the year before, I had seen a paper by David Cowen and Edwin Olmstead, neuropathologists at Columbia University, describing a primary axonal disease in infants: here the disease might present itself as early as the second year and was usually fatal by the age of seven. But in contrast to Hallervorden-Spatz disease, where the axonal abnormalities were confined to small but crucial areas, in infantile neuroaxonal dystrophy (as Cowen and Olmstead called it), there was widespread distribution of axonal swellings and fragments.

I wondered whether there were any animal models of axonal dystrophy, and here, by chance, I found that two researchers in our own neuropathology department were working on precisely this.[4] One of them, Stirling Carpenter, was working with rats maintained on a diet deficient in vitamin E; these pathetic rats lost control of their hind limbs and tails because sensation from these was blocked by axonal damage in the sensory tracts

---

4. It was, of course, more than a chance coincidence. A paper published in 1963 had described axonal changes associated with vitamin E deficiency in rats, and another paper in 1964 had described similar axonal changes in mice given IDPN (iminodipropionitrile). New findings have to be duplicated by other laboratories, and this is what my colleagues at UCLA were doing.

of the spinal cord, and their nuclei in the brain—a distribution of axonal damage completely unlike that seen in Hallervorden-Spatz disease but perhaps casting some light on the pathogenic mechanism involved.

Another colleague at UCLA, Anthony Verity, was working on an acute neurological syndrome that could be produced in laboratory animals by giving them a toxic nitrogen compound— iminodipropionitrile (IDPN).[5] Mice given this developed a wild excitement, incessantly turning in circles or running backwards, accompanied by involuntary jerking movements, bulging eyes, and priapism; they, too, had gross axonal changes, but these were in the arousal systems of the brain.

The term "waltzing mice" was sometimes used of such driven and incessantly active mice, but this decorous term gives no picture of the overwhelming severity of the syndrome. The usual quiet of the neuropathology department was punctuated by shrill cries and squeaks from the overexcited mice. While the IDPN-toxic mice were so different from the vitamin-E-deficient rats dragging their spread-eagled hind limbs behind them, and so different from the human conditions of Hallervorden-Spatz disease and infantile neuroaxonal dystrophy, they all seemed to share a common pathology: severe damage confined to the axons of nerve cells.

Could one get any insight from the fact that human and animal syndromes so different were generated, seemingly, by the same sort of axonal dystrophy, albeit in different regions of the nervous system?

---

5. IDPN and related compounds produce excessive arousal and hyperactivity not only in mammals but in fish, grasshoppers, and even protozoa.

After I moved to Los Angeles, I missed my Sunday morning rides to Stinson Beach with my motorcyclist friends,
and I reverted to being a lone rider again; on weekends, I would
embark on enormous solo rides. As soon as I could get away
from work on Friday, I saddled my horse—I sometimes thought
of my bike as a horse—and would set out for the Grand Canyon, five hundred miles away but a straight ride on Route 66.
I would ride through the night, lying flat on the tank; the bike
had only 30 horsepower, but if I lay flat, I could get it to a
little over a hundred miles per hour, and crouched like this, I
would hold the bike flat out for hour after hour. Illuminated
by the headlight—or, if there was one, by a full moon—the silvery road was sucked under my front wheel, and sometimes
I had strange perceptual reversals and illusions. Sometimes I
felt that I was inscribing a line on the surface of the earth, at
other times that I was poised motionless above the ground, the
whole planet rotating silently beneath me. My only stops were
at gas stations, to fill the tank, to stretch my legs and exchange
a few words with the gas attendant. If I held the bike at its
maximum speed, I could reach the Grand Canyon in time to
see the sunrise.

I would sometimes pull up at a little motel some distance
from the canyon where I would grab some sleep, but usually
I slept outside in my sleeping bag. There were sometimes
hazards to this—and not just bears or coyotes or insects. One
night, taking Route 33, the desert road from Los Angeles to San
Francisco, I stopped and unrolled my sleeping bag on what, in
the darkness, seemed to be a natural bed of beautiful soft moss.
Breathing in the clean desert air, I slept very well, but in the
morning I realized that I had bedded down on a huge mass of

fungal spores, which I must have been inhaling all through the night. It was *Coccidiomyces,* a notorious fungus native to the Central Valley that can cause anything from a mild respiratory illness to so-called Valley fever and, on occasion, a fatal pneumonia or meningitis. I developed a positive skin test for the fungus but, fortunately, no symptoms.

I would spend my weekends hiking in the Grand Canyon or sometimes in Oak Creek Canyon, with its marvelous red and purple colors. Sometimes I would go to Jerome, a ghost city (it was only years later that it was dolled up for tourists), and once I visited the grave of Wyatt Earp—one of the great romantic figures of the Old West.

I would ride back to Los Angeles on Sunday night and, with the resilience of youth, appear bright and fresh at neurology rounds at eight o'clock on Monday morning, with hardly a sign that I had ridden a thousand miles over the weekend.

•

Some people, perhaps more in the States than in Europe, have a "thing" against motorcycles and motorcyclists—a phobia or irrational hatred which may goad them into action.

My first experience of this was in 1963, when I was riding along Sunset Boulevard at a leisurely pace, enjoying the weather—it was a perfect spring day—and minding my own business. Seeing a car behind me in my driving mirror, I motioned the driver to overtake me. He accelerated, but when he was parallel with me, he suddenly veered towards me, making me swerve to avoid a collision. It didn't occur to me that this was deliberate; I thought the driver was probably drunk or incompetent. Having overtaken me, the car then slowed down. I slowed, too, until he motioned me to pass him. As I did so, he swung into the middle of the road, and I avoided being sideswiped by the narrowest margin. This time there was no mistaking his intent.

I have never started a fight. I have never attacked anyone unless I have been attacked first. But this second, potentially murderous attack enraged me, and I resolved to retaliate. I kept a hundred yards or more behind the car, just out of his line of sight, but prepared to leap forward if he was forced to stop at a traffic light. This happened when we got to Westwood Boulevard. Noiselessly—my bike was virtually silent—I stole up on the driver's side, intending to break a window or score the paintwork on his car as I drew level with him. But the window was open on the driver's side, and seeing this, I thrust my fist through the open window, grabbed his nose, and twisted it with all my might; he let out a yell, and his face was all bloody when I let go. He was too shocked to do anything, and I rode on, feeling I had done no more than his attempt on my life had warranted.

A second such incident happened when I was driving to San Francisco along the rarely traveled desert road, Route 33; I loved the emptiness of the road, the absence of traffic, and I was idling along at seventy miles an hour when a car appeared in my rearview mirror, going (I judged) near ninety. The driver had the whole road to overtake me but (like the driver in Los Angeles) tried to drive me off the road. He succeeded, and I was forced onto the shoulder reserved for emergencies and breakdowns, a soft shoulder. By a sort of miracle, I managed to hold the bike upright, throwing up a huge cloud of dust, and regained the road. My attacker was now a couple of hundred yards ahead. Rage more than fear was my chief reaction, and I snatched a monopod from the luggage rack (I was very keen on landscape photography at the time and always traveled with camera, tripod, monopod, etc., lashed to the bike). I waved it round and round my head, like the mad colonel astride the bomb in the final scene of *Dr. Strangelove*. I must have looked crazy—and dangerous—for the car accelerated. I accelerated too, and pushing the engine as much as I could, I started to overtake it. The driver tried to throw me off by driving erratically, suddenly slowing, or switching from side to side of the empty road, and when that failed, he took a sudden side road in the small town of Coalinga—a mistake, because he got into a maze of smaller roads with me on his tail and finally got trapped in a cul-de-sac. I leapt off the bike (all 260 pounds of me) and rushed towards the trapped car, waving the monopod. Inside the car I saw two teenage couples, four terrified people, but when I saw their youth, their helplessness, their fear, my fist opened and the monopod fell out of my hand.

I shrugged my shoulders, picked up the monopod, walked back to the bike, and motioned them on. We had all, I think, had the fright of our lives, felt the nearness of death, in our foolish, potentially fatal duel.

•

As I roved around California on my motorbike, I always carried my Nikon F with a range of lenses. I especially liked the macro lens, which allowed me to do close-ups of flowers and bark, lichens and moss. I also had a 4 x 5 Linhof view camera with a sturdy tripod. All this, packed in my sleeping bag, was well protected against bumps and jars.

I had known the magic of developing and printing photographs as a boy, when my little chemistry lab with its blackout curtains could serve as a darkroom, and I was to know it again at UCLA; we had a beautifully equipped darkroom in the neuropathology department, and I loved seeing images appear bit by bit as I rocked the large prints in the developing trays. Landscape photography was my favorite, and my weekend bike rides were sometimes inspired by *Arizona Highways*, which had marvelous photographs by Ansel Adams, Eliot Porter, and others—photographs which became my ideal.

I got an apartment near Muscle Beach in Venice, just south of Santa Monica. Muscle Beach had many greats, including Dave Ashman and Dave Sheppard, who had both lifted in the Olympic Games. Dave Ashman, a cop, had a modesty and sobriety very much the exception in a world of health nuts, steroids takers, drinkers, and braggarts. (Although I was taking plenty of other drugs in those days, I never took steroids myself.) I was told he was unmatched at the front squat, a much harder and trickier lift than the back squat, because one is holding the bar in front of one's chest rather than across one's shoulders, and one must maintain perfect balance and erectness. When I went one Sunday afternoon to the lifting platform on Venice Beach, Dave looked at me, the new kid, and challenged me to match him in the front squat. I could not decline the challenge; this would have branded me a weakling or a coward. I said, "Fine!" in what was meant to be a strong, confident voice but came out as a feeble croak. I matched him pound for pound, up to 500, but thought I was finished when he went from 500 to 550. To my surprise—I had hardly ever done front squats before—I matched him. Dave said that was his limit, but I, with a vainglorious impulse, asked for 575. I did this—just—though I had a feeling my eyes were bulging and wondered fearfully about the blood pressure in my head. After this, I was accepted on Muscle Beach and given the nickname Dr. Squat.

There were many other strong men on Muscle Beach. Mac Batchelor, who owned a bar to which we all flocked, had the largest and strongest hands I had ever seen; he was the world's undisputed arm-wrestling champion, and it was said that he could bend a silver dollar with his hands, though I never saw this. There were two gigantic men—Chuck Ahrens and Steve Merjanian—who had a semidivine status and were somewhat aloof from the rest of the Muscle Beach crowd. Chuck could do a one-arm side press with a 375-pound dumbbell, and Steve had invented a new lift—the incline bench press. Each of them weighed close to three hundred pounds and sported massive arms and chests; they were inseparable companions and completely filled the VW Beetle they shared.

Huge though he was, Chuck was eager to become even huger, and one day he appeared suddenly, filling the entire doorway, while I was working in neuropathology at UCLA. He had been wondering, he said, about human growth hormone—could I show him where the pituitary gland was located? I was surrounded by pickled brains, and I pulled one out of its jar to show Chuck the pea-size pituitary at the base of the brain. "So that's where it is," said Chuck, and, satisfied, took his leave. But I was disquieted: What was he thinking of? Had I been right in showing him the pituitary? I had fantasies of his raiding the neuropathology lab, going to the brains—a little formalin would not deter him—and plucking out their pituitaries, as one might pluck blackberries, and, even more gruesome, of his initiating a string of bizarre murders, in which the victims' heads would be cracked open, the brains torn out, and the pituitaries devoured.

Then there was Hal Connolly, an Olympic hammer thrower whom I often saw in Muscle Beach Gym. One of Hal's arms was almost paralyzed and hung loosely from his shoulder in a "waiter's tip" posture. The neurologist in me instantly recognized this as an Erb's palsy; such palsies come from traction on the brachial plexus during delivery if, as sometimes happens, a baby presents sideways and has to be pulled out by an arm. But if one of Hal's arms was useless, the other was a world-beater. His athleticism was a moving lesson in the power of will and compensation; it reminded me of what I sometimes saw at UCLA—patients with cerebral palsy and little use of their arms who had learned to write or play chess with their feet instead.

I took photographs on Muscle Beach, trying to catch its many characters and their haunts; this went hand in hand with a project for a book about the beach—descriptions of people and places, scenes and events, in that strange world which was Muscle Beach in the early 1960s.

Whether or not I *could* have written such a book, a montage of descriptions and verbal portraits interlarded with photographs, I do not know. When I left UCLA, I packed all my photographs, everything I had taken between 1962 and 1965, along with my sketches and notes, in a large suitcase. The suitcase never arrived in New York; no one seemed to know what had happened to it at UCLA, nor could I get an answer from post offices in L.A. or New York. So I lost almost all the photographs I had taken in my three years near the beach; only a dozen or so somehow survived. I like to imagine that the suitcase still exists and that it may turn up one day.

·

At Oxford, ca. 1953

With some of my fellow medical students at the Central Middlesex Hospital in 1957

With my new 250cc Norton motorbike in 1956

On a trip to Jerusalem in 1955, my mother greeted the future prime minister, Levi Eshkol. My father and I are standing behind her.

May 1961, on the road with Mac and Howard, my trucker companions

Lifting weights as a novice at the Maccabi club in London, 1956

I am standing at the left, taking in the scene on the lifting platform at Venice Beach.

Dr. Oliver Sacks of the Mar-Vel Athletic Club of San Rafael holds the new California State record in weightlifting.

At the Pacific Coast Championships in San Francisco last Saturday, the British medical doctor interning at Mount Zion Hospital in San Francisco, performed a full squat with 600 pounds across his shoulders.

A full squat with 600 pounds, a California state record I set in 1961

My official portrait as a
UCLA resident, and in the
neuropathology lab, 1964

My little house in
Topanga Canyon
was dwarfed by
an oak tree, but
was big enough to
accommodate a
piano.

Auntie Len

My mother

Thom Gunn, around the time we met in 1961

I took this photograph of my friend Carol
Burnett in Central Park in 1966.

Some of my own photographs, ca. 1963: a little shop in Topanga Canyon, Mel walking near Venice Beach, and a pool hall in Santa Monica

In New York, ca. 1970

At Muscle Beach with my beloved BMW motorbike

In Greenwich Village, 1961, on my new BMW R60

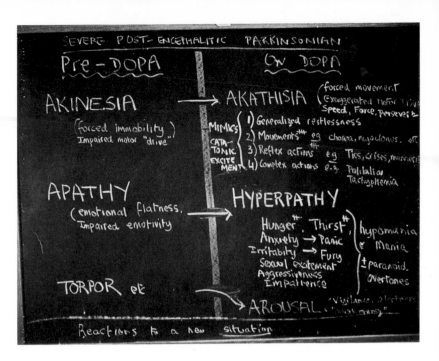

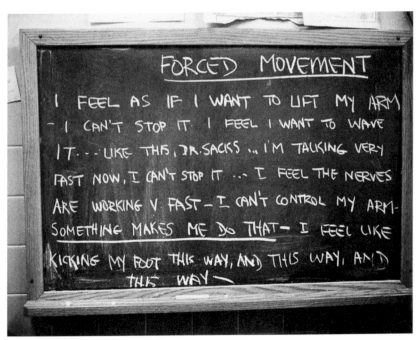

Some of my many "think boards" from *Awakenings* days

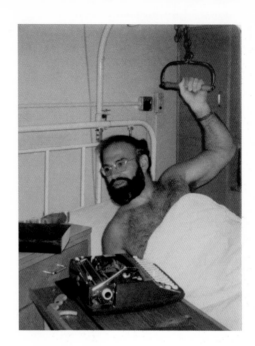

Recovering from
my leg accident in
Norway, 1974

Writing: on a car roof, in the Amsterdam train station, on a train

At Blue Mountain Center, 2010

Seeing patients at Bet Abraham, ca. 1988

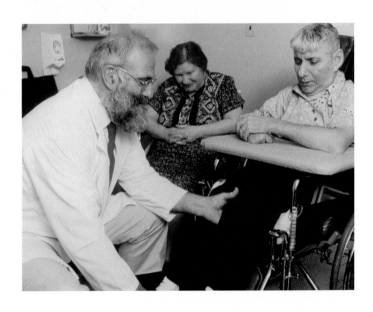

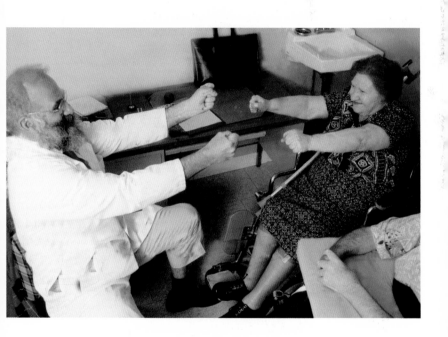

With Peter Brook and our Tourettic friend Shane Fistell, 1995

With Temple Grandin, 1994

With Robin Williams in 1989, on the set of *Awakenings*

Roger Hanlon and I share a love for squid, cuttlefish, and other cephalopods.

A still from the 1974 documentary of *Awakenings*

I tacked up a sign in my house on City Island reminding myself to say no to invitations so I could preserve writing time.

With my three brothers, David, Marcus, and Michael, at our parents' golden wedding anniversary in 1972

With my father at 37 Mapesbury on his ninety-second birthday in 1987

In Florence, 1988, where I had a revelatory dinner with Gerald M. Edelman

Talking with Edelman at another conference a few years later, in Bologna

Working with the Little Sisters of the Poor, 1995

Ralph Siegel, Bob Wasserman, Semir Zeki, and I presented a poster on the colorblind painter at the 1992 Society for Neuroscience annual meeting.

Strolling in 2010 on Darwin's Sandwalk with my
oldest friend, Eric Korn

Visiting Jonathan Miller in 1987 at his house in London

With Kate Edgar, my assistant and collaborator
for more than thirty years, in 1995

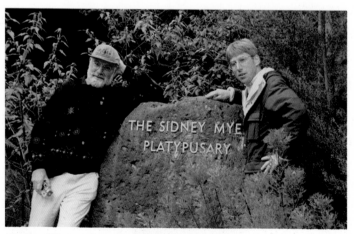

With Ralph Siegel at the "Platypusary" near Melbourne

I am happier in or under the water than on land: Snorkeling and walking the beach in Curaçao, emerging from Lake Tahoe, in scuba gear.

Receiving the award of Commander of the Order of the British Empire from Queen Elizabeth

In front of a favorite cycad at the Hortus Botanicus in Amsterdam, 2014

With Billy Hayes, 2014

Writing in my journal at Machu Picchu, 2006

Jim Hamilton was one of the weight-lifting crowd on Muscle Beach, but he was very different from the others. He had a huge head of curly hair and a huge curly beard and moustache, so very little of his face was visible except for the tip of his nose and his laughing, deep-set eyes. He was broad, barrel-chested, with a belly of Falstaffian proportions; he was one of the best bench pressers on the beach. He walked with a limp; one of his legs was shorter than the other and had surgical scars along its entire length. He had been in a motorcycle accident, he told me, sustaining multiple compound fractures, and had been in hospital for more than a year. He was eighteen at the time, just out of high school. It was a very difficult, lonely, pain-filled time, which would have been unbearable had he not discovered, to his own surprise and everyone else's, that he had a remarkable mathematical gift. This had not shown itself in school, which he had disliked, but now his only demands were for books on mathematics and game theory. The eighteen months of enforced physical inactivity—he had had nearly a dozen operations to reconstruct his shattered leg—was a time of great and exciting mental activity, as he moved with ever-greater power and freedom in the universe of mathematics.

Jim had had no idea what he would "do" when he graduated from high school, but when he left the hospital, his mathematical abilities got him a job as a computer programmer with the Rand Corporation. Few of his friends and drinking companions on Muscle Beach had any idea of Jim's mathematical side.

Jim had no fixed address; looking through our correspondence from the 1960s, I find postcards emanating from motels in Santa Monica, Van Nuys, Venice, Brentwood, Westwood, Hollywood, and dozens of other places. I have no idea what address he had on his driver's license; I suspect it may have been his boyhood address in Salt Lake City. He came from a distinguished Mormon family, descendants of Brigham Young.

It was easy for Jim to move from motel to motel or to sleep in his car, because he kept his few possessions—clothes and books, mostly—at Rand and sometimes spent the night there. He devised a variety of chess-playing programs for its super-computers, and he would test them (and himself) by playing chess with them. He especially enjoyed this when he was high on LSD; he felt that this made his games more unpredictable, more inspired.

If Jim had a circle of friends on Muscle Beach, he had another circle, of fellow mathematicians, and like the famous Hungarian mathematician Paul Erdös he might drop in on them in the middle of the night, spend a couple of hours brainstorming, and then stay over on their sofas for the rest of the night.

Before I met him, Jim spent occasional weekends in Las Vegas observing the blackjack tables, and he devised a strategy that would allow a player a slow but steady success at the game. Securing a three-month leave of absence from Rand, he ensconced himself in a hotel room in Las Vegas and played blackjack through his waking hours. Gradually but steadily, he won and accumulated more than $100,000, a very substantial amount in the late 1950s, but at this point he was visited by two large gentlemen. They said that his steady winning had been noted—he must have a "method" of some sort—but now it was time for him to leave town. Jim took their point and cleared out of Vegas the same day.

Jim had a huge, dirty once-white convertible at the time, filled with empty milk cartons and other rubbish; he drank a gallon or more of milk every day as he was driving and just tossed the empty cartons behind him. He and I took to each other in the muscle crowd. I liked to get Jim onto his own special passions—mathematical logic, game theory, and computer games—and he drew me out on my own interests and passions. When I got my little house in Topanga Canyon, he and his girlfriend, Kathy, would often visit.

·

As a neurologist, I was professionally interested in brain states, mind states, of all sorts, not least those induced or modified by drugs. New knowledge about psychoactive drugs and their effects on the brain's neurotransmitters was rapidly accumulating in the early 1960s, and I longed to experience these myself. Such experiences, I thought, might help me understand what some of my patients were going through.

Some of my Muscle Beach friends had urged me to try getting stoned on Artane, which I knew only as an anti-parkinsonian drug. "Just take twenty tablets," they said, "you'll still be in partial control. You'll find it a very different experience." So one Sunday morning, as I described in *Hallucinations*,

I counted out twenty pills, washed them down with a mouthful of water, and sat down to await the effect. . . . I had a dry mouth, large pupils, and found it difficult to read, but that was all. There were no psychic effects whatever—most disappointing. I did not know exactly what I had expected, but I had expected *something*.

I was in the kitchen, putting on a kettle for tea, when I heard a knocking at my front door. It was my friends Jim and Kathy; they would often drop round on a Sunday morning. "Come in, door's open," I called out, and as they settled themselves in the living room, I asked, "How do you like your eggs?" Jim liked them sunny side up, he said. Kathy preferred them over easy. We chatted away while I sizzled their ham and eggs—there were low swinging doors between the kitchen and the living room, so we could hear each other easily. Then, five minutes later, I shouted, "Everything's ready," put their ham and eggs on a tray, walked into the living room—and found it completely empty. No Jim, no Kathy, no sign that they had ever been there. I was so staggered I almost dropped the tray.

It had not occurred to me for an instant that Jim and Kathy's voices, their "presences," were unreal, hallucinatory. We had had a friendly, ordinary conversation, just as we usually had. Their voices were the same as always; there had been no hint, until I opened the swinging doors and found the living room empty, that the whole conversation, at least their side of it, had been completely invented by my brain.

I was not only shocked, but rather frightened, too. With LSD and other drugs, I knew what was happening. The world would look different, feel different; there would be every characteristic of a special, extreme mode of experience. But my "conversation" with Jim and Kathy had no special quality; it was entirely commonplace, with nothing to mark it as a hallucination. I thought about schizophrenics conversing with their "voices," but typically the voices of schizophrenia are mocking or accusing, not talking about ham and eggs and the weather.

"Careful, Oliver," I said to myself. "Take yourself in hand. Don't let this happen again." Sunk in thought, I slowly ate my ham and eggs (Jim and Kathy's, too) and then decided to go down to the beach, where I would see the real Jim and Kathy and all my friends, and enjoy a swim and an idle afternoon.

•

Jim was very much part of my life in Southern California—we would see each other two or three times a week—and I missed him acutely when I moved to New York. After 1970, his interest in computer games (including war games) spread to the use of computer animation in science-fiction films and cartoons, which kept him in Los Angeles.

When Jim visited me in New York in 1972, he looked well and happy; he was looking forward to the future, though it was not clear whether this would be in California or South America (he had spent a couple of years in Paraguay, where he greatly enjoyed life and had bought a ranch).

He said that he had not taken a drink in two years, and this especially pleased me because he had had a dangerous habit of going on sudden drinking binges, and the last one I knew about had given him pancreatitis.

He was on his way to Salt Lake City to spend time with his family. Three days later, I got a phone call from Kathy telling me that Jim was dead: he had gone on another drinking binge and again developed pancreatitis, but this time it was followed by necrosis of the pancreas and generalized peritonitis. He was only thirty-five.[6]

One day in 1963, I went bodysurfing off Venice Beach; it was rather rough water and no one else was in, but I, at the height of my strength (and grandiosity), was sure I could handle

---

6. I had hoped that it might be possible to publish some of Jim's mathematical work posthumously; I envisaged something like F. P. Ramsey's posthumous book *The Foundations of Mathematics* (Ramsey died at twenty-six). But Jim was essentially an on-the-spot problem solver: he would scribble an equation or a formula or a logical diagram on the back of an envelope, then crumple it up or lose it.

it. I was thrown around a bit—this was fun—but then a huge wave came towering far above my head. When I attempted to dive under it, I got flung on my back and tumbled over and over helplessly. I did not realize how far the wave had carried me until I saw it was about to crash me on the shore. Such impacts are the commonest cause of broken necks on the Pacific coast; I had just time to stick out my right arm. The impact tore my arm back and dislocated my shoulder, but it saved my neck. With one arm disabled, I could not crawl out of the surf quickly enough to get away from the next huge breaker, which was following close on the first. But at the last second, strong arms seized me and pulled me to safety. It was Chet Yorton, a very strong young bodybuilder. Once I was safely on the beach, in excruciating pain with the head of my humerus sticking out in the wrong place, Chet and some of his weight-lifting buddies grabbed me—two round the waist, two pulling the arm—until the shoulder went back into place with a squelch. Chet went on to win the Mr. Universe competition, and he was still superbly muscled at the age of seventy; I would not be here had he not pulled me out of the water in 1963.[7] The moment the joint was back in place, the shoulder pain vanished, and I became conscious of other pains in my arms and chest. I got

---

7. I should have learned that high seas are not for me, that a particular danger attended especially high "freak" waves, and that these can come out of the blue, even in seemingly calm seas. I later had two somewhat similar incidents. One was on Westhampton Beach on Long Island; this tore off the greater part of my hamstring muscles on the left side, and here again a friend, my old friend Bob Wasserman, pulled me to safety. Another one, which I was lucky to survive, happened when I was backstroking, foolishly, in high seas off the Pacific coast of Costa Rica.

I now have a dread of surf and have turned to lakes and slow-moving rivers as my favorite places to swim, though I still love snorkeling and scuba diving, which I learned in the calm waters of the Red Sea in 1956.

on my motorbike and rode to the emergency room at UCLA, where they found I had a broken arm and several broken ribs.

·

There were occasional weekends when I was on call at UCLA and others when I supplemented my meager income by moonlighting at the Doctors Hospital in Beverly Hills. On one occasion there, I met Mae West, who was in for some small operation. (I did not recognize her face, for I am face-blind, but I recognized her voice—how could one not?) We chatted a good deal. When I came to say good-bye to her, she invited me to visit her in her mansion in Malibu; she liked to have young musclemen around her. I regret that I never took up her invitation.

Once my own strength came in handy on the neurological wards. We were testing visual fields in a patient unlucky enough to have developed a coccidiomyces meningitis and some hydrocephalus. While we were testing him, his eyes suddenly rolled up in his head and he started to collapse. He was "coning"; this is the rather mild term used for a terrifying event in which, with excessive pressure in the head, the cerebellar tonsils and brain stem get pushed through the foramen magnum at the base of the skull. Coning can be fatal within seconds, and with the speed of reflex I grabbed our patient and held him upside down; his cerebellar tonsils and brain stem went back into the skull, and I felt I had snatched him from the very jaws of death.

Another patient on the ward, blind and paralyzed, was dying from a rare condition called neuromyelitis optica, or Devic's disease. When she heard that I had a motorbike and lived in Topanga Canyon, she expressed a special last wish: she wanted to come for a ride with me on my motorbike, up and down the loops of Topanga Canyon Road. I came to the hospital one Sunday with three weight-lifting buddies, and we managed to abduct the patient and lash her securely to me on the back of the bike. I set out slowly and gave her the ride in Topanga she desired. There was outrage when I got back, and I thought I would be fired on the spot. But my colleagues—and the patient—spoke up for me, and I was strongly cautioned but not dismissed. In general, I was something of an embarrassment to the neurology department but also something of an ornament—the only resident who had published papers—and I think this might have saved my neck on several occasions.

•

I sometimes wonder why I pushed myself so relentlessly in weight lifting. My motive, I think, was not an uncommon one; I was not the ninety-eight-pound weakling of bodybuilding advertisements, but I was timid, diffident, insecure, submissive. I became strong—very strong—with all my weight lifting but found that this did nothing for my character, which remained exactly the same. And, like many excesses, weight lifting exacted a price. I had pushed my quadriceps, in squatting, far beyond their natural limits, and this predisposed them to injury, and it was surely not unrelated to my mad squatting that I ruptured one quadriceps tendon in 1974 and the other in 1984. While I was in hospital in 1984, feeling sorry for myself, with a long cast on my leg, I had a visit from Dave Sheppard, mighty Dave, from Muscle Beach days. He hobbled into my room slowly and painfully; he had very severe arthritis in both hips and was awaiting total hip replacements. We looked at each other, our bodies half-destroyed by lifting.

"What fools we were," Dave said. I nodded and agreed.

I liked him as soon as I saw him working out at the Central Y in San Francisco; it was early in 1961. I liked his name: Mel, Greek for "honey" or "sweet." As soon as he told me his name, a string of mel-words ran through my mind—"mellify," "melliferous," "mellifluous," "mellivorous" . . .

"Nice name—Mel," I said. "Mine's Oliver."

He had a husky, athlete's body, with powerful shoulders and thighs, and flawlessly smooth milk-white skin. He was all of nineteen years old, he told me. He was in the navy—his ship, the USS *Norton Sound*, was stationed in San Francisco—and he trained when he could at the Y. I too was training hard at the time, getting ready for what I hoped might be a record-breaking squat, and our workout times would sometimes coincide.

After a workout and shower, I would run Mel back to his ship on my motorbike. He had a soft brown deerskin jacket—he had shot the deer, he said, in his native Minnesota—and I gave him the spare crash helmet I always kept on the bike. I thought we made a good pair, and I had a little tickle of excitement at his sitting on the saddle behind me and holding me firmly around the waist; it was his first time on a motorcycle, he said.

We enjoyed each other's company for a year—the year of my internship at Mount Zion. We would go on weekend motorbike rides together, camping out, swimming in ponds and lakes, and sometimes wrestling together. There was an erotic frisson here for me, and perhaps for Mel too. Erotic with the urgent opposition of our bodies, though there was no explicit sexual element, nor would an observer have thought we were anything more than a couple of young men wrestling together. Both of us were proud of our washboard abdominals and would do sets of sit-ups, a hundred or more at a time. Mel would sit astride me, punching me playfully in the stomach with each sit-up, and I would do the same with him.

This I found sexually exciting, and I think he did too; Mel was always saying, "Let's wrestle," "Let's do abs," though it was not a purposively sexual act. We could work our abdominals or wrestle and get pleasure from it, at one and the same time. So long as things went no further.

I felt Mel's fragility, his not fully conscious, lurking fear of sexual contact with another man, but also the special feeling he had for me, which, I dared to think, might transcend these fears. I realized I would have to go very gently.

Our bucolic and in a sense innocent honeymoon, with delight in the present and little thought of the future, lasted a year, but as the summer of 1962 approached, we had to make plans.

Mel's navy service was coming to an end—he had gone straight into the navy from high school—and he hoped now to go to college. I was committed to moving to Los Angeles for my residency at UCLA, so we arranged to share an apartment in Venice, California—close to Venice Beach and Muscle Beach Gym, where we could train. I helped Mel with the application forms to Santa Monica College, and I bought him a secondhand BMW, a twin to my own machine. He did not like accepting gifts or money from me and secured a job for himself in a carpet factory a short walk from our apartment.

The apartment was small, a studio with a kitchenette. Mel and I had separate beds, and the rest of the apartment was filled with books and the ever-accumulating journals and papers I had been writing for years; Mel had very few possessions of his own.

Mornings were pleasant: we enjoyed coffee and breakfast together and then going our separate ways to work—Mel to the carpet factory, I to UCLA. After work, we would go down to Muscle Beach Gym and then to Sid's Café on the beach, where the muscle crowd hung out. Once a week we would go to a movie, and a couple of times a week Mel would take off on his own for a motorcycle ride.

Evenings could be a strain: I found it difficult to concentrate and was very conscious, almost hyperaware, of Mel's physical presence, not least his virile animal smell, which I loved. Mel liked being massaged and would lie naked facedown on his bed and ask me to massage his back. I would sit astride him, wearing my training shorts, and pour oil on his back—neat's-foot oil, which we used to keep our motorbike leathers supple—and slowly massage his shapely, powerful back muscles. He enjoyed this, relaxing and surrendering to my hands, and I enjoyed it too; indeed, it would bring me to the brink of orgasm. The brink was okay—just; one could pretend that nothing special was happening. But on one occasion, I could not contain myself and spurted semen all over his back. I felt him suddenly stiffen when this happened, and without a word he got up and had a shower.

He would not speak to me for the rest of the evening; it was evident that I had gone too far. (I suddenly thought of my mother's words and how *MEL* were my mother's initials—Muriel Elsie Landau.)

The next morning Mel said tersely, "I have to move out, find a place of my own." I said nothing but felt close to tears. He told me that some weeks before, on one of his evening motorcycle rides, he had met a young woman—not so young actually, she had a couple of teenage children—and she had invited him to stay in her house. He had put this off for the sake of friendship, but now, he felt, he had to leave me. He hoped, nonetheless, that we could remain "good friends."

I had not met her, but I felt she had taken Mel from me. I wondered, thinking of Richard ten years earlier, whether it was my fate to fall in love with "normal" men.

I felt desperately lonely and rejected when Mel moved out, and it was at this juncture that I turned to drugs, as some sort of compensation. I rented the little house in Topanga Canyon; it was rather isolated, being at the top of an unpaved trail, and I resolved never to live with anyone again.[8]

•

In fact, Mel and I kept in touch for another fifteen years, although there were always troubling undercurrents beneath the surface—more so, perhaps, with Mel, for he was not fully at ease with his own sexuality and longed for physical contact with me, where I had, so far as sex was concerned, given up my illusions and hopes about him.

Our last meeting was no less ambiguous. I was visiting San Francisco in 1978, and Mel arranged to come down from Oregon. He was curiously and uncharacteristically nervous and insisted we go to a bathhouse together. I had never been to a bathhouse; San Francisco's gay bathhouses were not to my taste. When we stripped off, I saw that Mel's skin, so milky and flawless before, was now covered with brownish "café au lait" patches. "Yes, it's neurofibromatosis," he said. "My brother has it too.

---

8. A few years later, Topanga Canyon was to become a mecca for musicians, artists, and hippies of all kinds, but in the early 1960s, when I was there, it was relatively unpopulated and very quiet. Houses on unpaved trails, like mine, had no close neighbors, and I had to get water trucked up in fifteen-hundred-gallon deliveries, to be stored in a cistern.

I thought you should see it," he added. I hugged Mel and wept. I thought of Richard Selig showing me his lymphosarcoma— were the men I loved fated to get terrible diseases? We said good-bye, shaking hands rather formally when we left the bath-house. We never met or wrote to each other again.

I had had dreams, in our "honeymoon" period, that we would spend our lives together, even into a happy old age; I was all of twenty-eight at the time. Now I am eighty, trying to reconstruct an autobiography of sorts. I find myself thinking of Mel, of us together, in those early, lyrical, innocent days, wondering what happened to him, whether he is still alive (neurofibromatosis, von Recklinghausen's disease, is an unpredictable animal). I wonder if he will read what I have just written and think more kindly of our ardent, young, very confused selves.

While I took Richard Selig's delicate disengagement ("I am not that way, but I appreciate your love and love you too, in my own way") without feeling rebuffed or broken-hearted, Mel's almost disgusted rejection affected me deeply, depriving me (so I felt) of all hope of a *real* love life, driving me inwards and downwards to seek whatever satisfactions I could find with drug-fueled fantasy and pleasure.

During my two years in San Francisco, I had engaged in a sort of harmless weekend duplicity, exchanging my white intern's jacket for animal skins and taking off on the bike, but now I was driven to a darker, more dangerous duplicity. From Monday to Friday, I devoted myself to my patients at UCLA, but those weekends I did not take off on my bike, I devoted to virtual travel—drug trips on cannabis, morning glory seeds, or LSD. These were secret, shared with no one, mentioned to no one.

One day a friend offered me a "special" joint; he did not say what was special about it. I took a puff, nervously, then another, and then, voraciously driven, smoked the rest, voraciously because it was producing what cannabis alone had never produced—a voluptuous, almost orgasmic feeling of great intensity. When I asked what the joint had contained, I was told that it had been doped with amphetamine.

I do not know how much a propensity to addiction is "hardwired" or how much it depends on circumstances or state of mind. All I know is that I was hooked after that night with an amphetamine-soaked joint and was to remain hooked for the next four years. In the thrall of amphetamines, sleep was impossible, food was neglected, and everything was subordinated to the stimulation of the pleasure centers in my brain.

It was while I was wrestling with amphetamine addiction—I had rapidly moved from speed-doped marijuana to methamphetamine by mouth or vein—that I read about James Olds's experiments with rats. The rats had electrodes implanted in the reward centers of their brains (the nucleus accumbens and other deep subcortical structures), and they were able to stimulate these centers by pressing a lever. They would do so nonstop, until they died of exhaustion. Once I was loaded with amphetamines, I felt as helplessly driven as one of Olds's rats. The doses I took got larger and larger, pushing my heart rate and my blood pressure to lethal heights. There was an insatiability in this state; one could never get enough. The ecstasy of amphetamines was mindless and all sufficient—I needed nothing and nobody to "complete" my pleasure—it was essentially complete, though completely empty. All other motives, goals, interests, desires, disappeared in the vacuousness of the ecstasy.

I gave little thought to what this was doing to my body and perhaps my brain. I knew a number of people on Muscle Beach and Venice Beach who had died from massive doses of amphetamines, and I was very lucky I did not get a heart attack or a stroke myself. I did and did not realize I was playing with death.

I would be back at work on Monday mornings—shaken and almost narcoleptic—but no one, I think, realized that I had been in interstellar space, or reduced to an electrified rat, over the weekend. When people asked what I had done over the weekend, I would say that I had been "away"—how "away," and in what sense, they probably did not guess.

·

By now, I had had a couple of papers published in neurology journals, but I hoped for something more—an exhibit at the upcoming annual meeting of the American Academy of Neurology (AAN).

With the help of the department's excellent photographer, Tom Dolan—a friend who shared my interests in marine biology and invertebrates—I turned from photography of western landscapes to the inner landscapes of neuropathology. We worked hard at getting the best possible pictures to convey the microscopic appearance of the hugely distended axons seen in Hallervorden-Spatz disease, vitamin E deficiency in rats, and IDPN intoxication in mice. We made these into large Kodachrome transparencies and constructed a special lighted viewing cabinet to illuminate the transparencies from within, along with captions to go with the photographs. It took months to put all this together, pack it up, and erect it for the 1965 spring meeting of the AAN in Cleveland. Our exhibit was a hit, as I hoped it might be, and I, normally shy and reticent, found myself drawing people over to the exhibit, expatiating on the special beauties and interests of our three axonal dystrophies, so different clinically and topographically yet so similar at the level of individual axons and cells.

The exhibit was my way of introducing myself, saying, "Here I am, look what I can do," to the neurological community in the States, just as my California squat record, four years earlier, had been my introduction to the weight-lifting community on Muscle Beach.

I had feared that I might be unemployed when my residency came to an end in June of 1965. But the axonal dystrophy exhibit led to job offers from all over the United States, including two especially prized ones from New York City—one from Cowen and Olmstead at Columbia University, and one from Robert Terry, a distinguished neuropathologist at the Albert Einstein College of Medicine. I had become enamored of Terry's pioneer work when he came to UCLA in 1964 to present his latest electron microscopic findings in Alzheimer's disease; at the time, my interests were especially in degenerative diseases of the nervous system, whether they occurred in youth, like Hallervorden-Spatz disease, or in old age, like Alzheimer's.

I could perhaps have stayed on at UCLA, living in my little house in Topanga Canyon, but I felt I needed to move on and specifically to go to New York. I felt that I was enjoying California too much, was getting addicted to an easy, sleazy life, to say nothing of a deepening drug addiction. I felt I needed to go to a hard, real place, a place where I could devote myself to work and perhaps discover or create a real identity, a voice of my own. Despite my interest in axonal dystrophy—Cowen and Olmstead's special field—I wanted to do something else, to combine neuropathology and neurochemistry in some intimate way. Einstein was a very new college which offered special interdisciplinary fellowships in neuropathology and neurochemistry—disciplines brought together by the genius of Saul Korey—and so I accepted the offer from Einstein.[9]

•

In my three years at UCLA, I worked hard, played hard, and took no holidays. Every so often, I would go to my chief, the formidable (but kindly) Augustus Rose, to say I wanted a few days off, but he would always reply, "Every day is a holiday for you, Sacks," and, cowed, I would drop the idea.

But I continued my weekend bike rides, and I often rode to Death Valley, sometimes to Anza-Borrego; I loved the desert. Occasionally, I rode down to Baja California, getting the feel of a completely different culture, though the road was very rough beyond Ensenada. I put more than 100,000 miles on my bike by the time I left UCLA and moved to New York. Roads were starting to get crowded by 1965, especially in the East, and I was never again to enjoy biking, life on the road, with the sort of freedom and joy I had had in California.

I sometimes wonder why I have spent more than fifty years in New York, when it was the West, and especially the Southwest, which so enthralled me. I now have many ties in New York—to my patients, my students, my friends, and my analyst—but I have never felt it move me the way California did. I suspect my nostalgia may be not only for the place itself but for youth, and a very different time, and being in love, and being able to say, "The future is before me."

---

9. Korey, a man of immense farsightedness, imagined the rise of a unified neuroscience years before the term was invented. I never met him, for he died, tragically young, in 1963, but he left as his legacy the close interaction of all the "neuro" labs (as well as the clinical neurology departments) at Einstein—an interaction that continues today.

Out of Reach

In September of 1965, I moved to New York to do my fellow-ship in neurochemistry and neuropathology at the Albert Einstein College of Medicine. I still had hopes of being a real scientist, a bench scientist, even though my research at Oxford had ended disastrously and should have warned against any repetitions. But with the jauntiness of denial, I thought I should have another try.

The fiery Robert Terry, whose beautiful electron microscope studies of Alzheimer's disease had so enthralled me when he spoke at UCLA, was away on a sabbatical when I arrived in New York, and in his absence the department of neuropathology was run by Ivan Herzog, a mild, sweet-tempered Hungarian émigré who was amazingly tolerant of and patient with his erratic fellow.

By 1966, I was taking very heavy doses of amphetamines, and I became—psychotic? manic? disinhibited? enhanced? I hardly know what term to use, but it went with an extraordinary heightening of the sense of smell and of my normally unremarkable powers of imagery and memory.

We would have a teaching quiz every Tuesday, when Ivan asked us to identify photomicrographs of unusual neuropathological conditions. I was normally very bad at this, but one Tuesday Ivan presented some photographs, saying, "This is an extremely rare condition—I don't expect you to recognize it."

I cried out, "Microglioma!" Everyone looked at me, startled. I was normally the mute one.

"Yes," I continued, "only six cases have been described in the world literature." And I cited all these in detail. Ivan stared at me, goggling.

"How do you know this?" he asked.

"Oh, just casual reading," I answered, but I was as startled as he was. I had no idea of how, or indeed when, I could have absorbed this knowledge so swiftly and unconsciously. It was all part of that strange amphetamine enhancement.

·

As a resident, I had been especially interested in rare, often familial diseases called lipidoses, in which abnormal fats accumulate in brain cells. I was excited when it was discovered that these lipids could also accumulate in the nerve cells residing in the wall of the gut. This might make it possible to diagnose such diseases even before any symptoms appeared, by doing a biopsy not of the brain but of the rectum—a far less traumatic procedure. (I had seen the original report of this in the *British Journal of Surgery*.) One needed to find only a single lipid-distended neuron to make a diagnosis. I wondered whether other diseases—Alzheimer's, for example—could also cause changes in the neurons of the gut and be diagnosed very early this way. I developed, or adapted, a technique for "clearing" the rectal wall so that it was almost transparent and staining the nerve cells with methylene blue; this way one could see dozens of nerve cells in a low-power microscope field, increasing the chances of finding any abnormalities. Looking at our slides, I persuaded myself and my chief, Ivan, that we could see changes in rectal nerve cells—the neurofibrillary tangles and Lewy bodies that seemed characteristic of Alzheimer's and of Parkinson's disease. I had high thoughts of the importance of our findings; it would be a breakthrough, an invaluable diagnostic technique. In 1967, we submitted an abstract of the paper we hoped to present at the upcoming meeting of the American Academy of Neurology.

Unfortunately, things went wrong at this point. We needed much more material besides the few rectal biopsies we had, but we were unable to obtain these.

We could not proceed with our research, and Ivan and I pondered the matter—should we retract our preliminary abstract? In the end, we did not, feeling that others would examine the matter; the future would decide. And so it did: the "discovery" which I had hoped would make my name as a neuropathologist turned out to be an artifact.

•

I had an apartment in Greenwich Village, and unless there was deep snow, I would ride my motorbike to work in the Bronx. I had no saddlebags, but the bike had a sturdy rack on the back to which I could secure what I needed with strong elastic bands.

My project in neurochemistry was to extract myelin, the fatty material which sheathes larger nerve fibers, enabling them to conduct nerve impulses more speedily. There were many open questions at the time: Was invertebrate myelin, if it could be extracted, different in structure or composition from vertebrate myelin? I chose earthworms as my experimental animal; I had always been fond of them, and they have giant, myelin-sheathed, rapidly conducting nerve fibers that permit sudden, massive movements when they are threatened. (It was for this reason that I had chosen earthworms to work with ten years before, to study the demyelinating effects of TOCP.)

I committed a veritable genocide of earthworms in the college garden: thousands of earthworms would be needed to extract a respectable sample of myelin; I felt like Marie Curie processing her tons of pitchblende to obtain a decigram of pure radium. I became adept at dissecting out the nerve cord and cerebral ganglia in a single, swift excision, and I would mash these up to make a thick, myelin-rich soup ready for fractionation and centrifugation.

I kept careful notes in my lab notebook, a large green volume which I sometimes took home with me to ponder over at night. This was to prove my undoing, for, rushing to get to work one morning after oversleeping, I failed to secure the elastic bands on the bike rack and my precious notebook, containing nine months of detailed experimental data, escaped from the loose strands and flew off the bike while I was on the Cross Bronx Expressway. Pulling over to the side, I saw the notebook dismembered page by page by the thunderous traffic. I tried darting into the road two or three times to retrieve it, but this was madness, for the traffic was too dense and too fast. I could only watch helplessly until the whole book was torn apart.

I consoled myself, when I got to the lab, by saying at least I have the myelin itself; I can analyze it, look at it under the electron microscope, and regenerate some of the lost data. Over the ensuing weeks, I managed to do some good work and had started to feel some optimism again, despite some other mishaps, as when, in the neuropathology lab, I screwed the oil-immersion objective of my microscope through several irreplaceable slides.

Even worse, from my bosses' point of view, I managed to get crumbs of hamburger not only on my bench but in one of the centrifuges, an instrument I was using to refine the myelin samples.

Then a final and irreversible blow hit me: I *lost* the myelin. It disappeared somehow—perhaps I swept it into the garbage by mistake—but this tiny sample which had taken ten months to extract was irretrievably gone.

A meeting was convened: no one denied my talents, but no one could gainsay my defects. In a kindly but firm way, my bosses said to me, "Sacks, you are a menace in the lab. Why don't you go and see patients—you'll do less harm." Such was the ignoble beginning of a clinical career.[1]

A ngel dust—what a sweet, alluring name! Deceptive, too, for its effects can be very far from sweet. A rash drug taker in the 1960s, I was prepared to try almost anything, and knowing my insatiable, dangerous curiosity, a friend invited me to join an angel dust "party" in an East Village loft.

I arrived a bit late—the party had begun—and when I opened the door, I was faced by a scene so surreal, so insane, that it made the Mad Hatter's tea party seem, by comparison, the essence of sanity and propriety. There were almost a dozen people there, all of them flushed, some with bloodshot eyes, several staggering. One man was uttering shrill cries and leaping about the furniture; perhaps he fancied himself a chimp.

---

1. Perhaps I had never really expected to succeed in research. In a 1960 letter to my parents, wondering about doing research in physiology at UCLA, I wrote, "I am probably too temperamental, too indolent, too clumsy and even too dishonest to make a good research worker. The only things I really enjoy are talking . . . reading and writing."

And I quoted a letter I had just received from Jonathan Miller, who, writing about himself, Eric, and me, said, "I am, like Wells, enchanted by the prospect and paralyzed by the reality of scientific research. The only place where any of us move nimbly or with grace is with ideas and words. Our love of science is utterly literary."

Another was "grooming" his neighbor, picking imaginary insects off his arms. One had defecated on the floor and was playing with the feces, making patterns in it with an index finger. Two of the guests were motionless, catatonic, and another was making faces and blathering, a farrago which sounded like schizophrenic "word salad." I phoned emergency services, and all of the partiers were taken to Bellevue; some of them had to be hospitalized for weeks. I was exceedingly glad I had arrived late and had not taken any angel dust myself.

•

Later, working as a neurologist at Bronx State, I saw a number of people who had been precipitated by angel dust (phencyclidine, or PCP) into schizophrenia-like states sometimes lasting months on end. Some had seizures, too, and many of them, I found, had highly abnormal EEGs for as much as a year after taking the angel dust. One of my patients murdered his girlfriend when they were on PCP, though he retained no memory of the deed. (Years later I published an account of this very complex and tragic business, and its equally complex and tragic sequelae, in *The Man Who Mistook His Wife for a Hat.*)

PCP was originally introduced as an anesthetic in the 1950s but by 1965 was no longer in medical use because of its horrific side effects. Most hallucinogens have their primary effects on serotonin, one of the brain's many neurotransmitters, but PCP, like ketamine, impairs the transmitter glutamine and is far more dangerous and long lasting in effect than other hallucinogens. It is known to cause structural lesions, as well as chemical changes, in the brains of rats.[2]

_____

2. One might imagine that the recreational use of angel dust would not have outlasted the 1960s, but I find, looking at the latest available Drug Enforcement Administration figures, that as recently as 2010 more than

The summer of 1965 was a particularly difficult, dangerous time for me, because I had three structureless months on my hands between finishing at UCLA and starting at Einstein.

I sold my trusty BMW R60 and went to Europe for a few weeks, where I bought a new, more modest model, an R50, at the BMW works in Munich. I went first to the little village of Gunzenhausen near Munich to see the graves of some of my ancestors; some of them were rabbis and had taken the name Gunzenhausen for themselves.

Then I went to Amsterdam, which had long been my favorite city in Europe and where I had had my sexual baptism, my introduction to gay life, ten years before. I had met a number of people on previous visits, and now, at a dinner party, I met a young German theater director called Karl. He was elegantly dressed and articulate; he spoke with wit and knowledge about Bertolt Brecht, many of whose plays he had directed. I thought him charming and civilized but did not think of him as especially attractive in sexual terms and gave no thought to him when I returned to London.

I was surprised, therefore, to get a postcard from him a couple of weeks later, suggesting that we meet in Paris. (My mother saw the postcard, asked whom it came from, and might have been mildly suspicious; I said, "An old friend," and we left it at that.)

The invitation intrigued me, so I went to Paris by road and ferry, taking my new motorbike with me. Karl had found a comfortable hotel room with a commodious double bed. We

---

fifty thousand young adults and high school students had to be brought to emergency rooms after taking PCP.

divided our long weekend in Paris between sightseeing and lovemaking. I had brought a cache of amphetamines with me and downed twenty tablets or so before we went to bed. Hot with excitement and desire, which I had not felt before taking the tablets, I made love ardently. Karl, surprised by my ardor and insatiability, asked what had got into me. Amphetamines, I said, and showed him the bottle. Curious, he took one, liked the effect, took another, and another, and another, and soon, like me, he was fizzing with the stuff, as excited as if he had been in Woody Allen's "orgasmatron." After I know not how many hours, we separated, exhausted, had a little break, and started back again.

That we might thrash about like two animals in rut was perhaps not wholly surprising, given the circumstances plus the amphetamine. But what I did not expect was that this experience would cause us to fall in love with each other.

When I got back to New York in October, I wrote fevered love letters to Karl and received equally fervent ones in return. We idealized each other; we saw ourselves spending long, loving, creative lives together—Karl fulfilling himself as an artist, I as a scientist.

But then the feeling started to fade. We asked ourselves whether the experience we had shared was real, authentic, given the huge aphrodisiac thrust of the amphetamines. I found this question particularly humiliating—could so lofty a transport as falling in love be reduced to something purely physiological?

In November, we oscillated between doubt and affirmation, thrown from one pole to the other. By December, we were out of love and (without regretting or denying the strange fever which had gripped us) had no desire to continue our correspondence. In my last letter to him, I wrote, "I have memories of a fevered joy, intense, irrational . . . totally gone."

Three years later, I received a letter from Karl, telling me he would be coming to live in New York. I was curious to see him, to reencounter him, now that I had ceased to take drugs.

He had got a small apartment on Christopher Street near the river, and when I entered, I found the air ill-smelling and thick with smoke. Karl himself, so elegant before, was unshaven, unkempt, unclean. There was a dirty mattress on the floor and pillboxes in a shelf above it. I saw no books, no residue of his past life as a reader and director. He seemed to have no interest in anything intellectual or cultural. He had become a drug dealer and would talk about nothing except drugs and how the world could be saved by LSD. His eyes had an opaque, fanatical look. I found all this bewildering and shocking. What had happened to the fine, gifted, civilized man I had met just three years earlier?

I had a sense of horror—and, in part, of guilt. Was it not I who had introduced Karl to drugs? Was I, to some extent, responsible for this shattering of a once-noble human being? I did not see Karl again. I heard, in the 1980s, that he was ill with AIDS and had gone back to Germany to die.

While I was doing my residency at UCLA, Carol Burnett, my friend from Mount Zion, had returned to New York for her residency in pediatrics. When I moved to New York, we resumed our friendship; we would often go to Barney Greengrass ("the Sturgeon King") for a smoked-fish brunch on Sunday mornings. Carol had grown up on the Upper West Side, had grown up going to Barney Greengrass, and she had picked up her fluent, idiomatic Yiddish from listening to the Yiddish nattering which filled the shop and restaurant on Sunday mornings.

In November of 1965, I was taking huge doses of amphetamines every day and then, unable to sleep, huge doses of chloral hydrate, a hypnotic, every night. One day, sitting in a café, I started to experience hallucinations of the wildest sort, which came on suddenly, as I described in *Hallucinations:*

> As I was stirring the coffee, it suddenly turned green, then purple. I looked up, startled, and saw that a customer paying his bill at the cash register had a huge proboscidean head, like an elephant seal. Panic seized me; I slammed a five-dollar note on the table and ran across the road to a bus on the other side. But all the passengers on the bus seemed to have smooth white heads like giant eggs, with huge glittering eyes like the faceted compound eyes of insects—their eyes seemed to move in sudden jerks, which increased the feeling of their fearfulness and alienness. I realized that I was hallucinating, that I could not stop what was happening in my brain, and that I had to maintain at least an external control and not panic or scream or become catatonic, faced by the bug-eyed monsters around me.

When I got off the bus, the buildings around me were tossing and flapping from side to side, like flags blowing in a high wind. I called Carol.

"Carol," I said, as soon as she picked up, "I want to say goodbye. I've gone mad, psychotic, insane. It started this morning, and it's getting worse all the while."

"Oliver!" Carol said. "What have you just taken?"

"Nothing," I replied. "That's why I'm so frightened." Carol thought for a moment, then asked, "What have you just *stopped* taking?"

"That's it!" I said. "I was taking a huge amount of chloral hydrate and ran out of it last night."

"Oliver, you chump! You always overdo things," Carol said. "You've given yourself the DTs."

Carol sat with me, nursed me, anchored me, through the four days of my delirium, when waves of hallucination and delusion kept threatening to engulf me; she was the only stable point in a chaotic and shattered world.

The second time I called her in a panic was three years later, when I started, one evening, to feel a little dizzy, light-headed, and strangely excited for no reason. I could not sleep and was alarmed by seeing little patches of my skin changing color before my eyes. My landlady at the time was a brave and charming old lady who had battled for years with scleroderma, a very rare disease which gradually hardens and shrinks the skin, causing deformities of the limbs and sometimes neces-sitating amputations. Marie had had this for more than fifty years; she told me proudly that she was the longest-surviving case known to the medical profession. In the middle of the night, when bits of my skin seemed to change texture, becom-ing hard and waxy, I had a sudden, piercing insight: I too had scleroderma, "galloping scleroderma." I had never actually heard of such a thing; scleroderma is usually the most indolent of diseases. But there is always a first case of this or that, and I thought I too would surprise the profession, as the world's first case of *acute* scleroderma.

I phoned Carol, and she came to see me, black bag in hand.

She took one look at me—I had a high fever and was covered in blisters—and said, "Oliver, you idiot, you've got chicken pox."

"Have you examined anyone with shingles lately?" she continued. Yes, I told her. I had examined an old chap at Beth Abraham with shingles just fourteen days earlier. "Experientia docet," Carol said. "Now you *know*, and not just because the textbooks say so, that shingles and chicken pox come from one and the same virus."

Brilliant, witty, generous Carol, fighting juvenile diabetes in herself, as well as prejudices against women and blacks in her profession, rose to become a dean at Mount Sinai and in this capacity was critical, for many years, in ensuring that women physicians and physicians of color were respected and treated equitably. She never forgot the episode with the surgeons at Mount Zion.

Drugging had increased when I started in New York, fueled in part by the soured love affair with Karl and in part by the fact that my work was going badly and my sense that I should not have opted for research in the first place. By December of 1965, I had started to call in sick, missing work for days at a time. I was taking amphetamines constantly and eating very little; I lost so much weight—nearly eighty pounds in three months—that I could scarcely bear the sight of my emaciated face in the mirror.

On New Year's Eve, I had a sudden lucid moment in the midst of an amphetamine ecstasy, and I said, "Oliver, you will not see another New Year's Day unless you get help. There has to be some intervention." I felt there were very deep psychological troubles underlying my addiction and self-destructiveness and that unless these were addressed, I would always return to drugs, and sooner or later do myself in.

A year or so earlier, when I was still in Los Angeles, Augusta Bonnard, a friend of the family's who was herself a psychoanalyst, had suggested that I go and see someone. I went reluctantly to see the psychoanalyst she recommended, a Dr. Seymour Bird. When he asked, "Well, what brings you here, Dr. Sacks?" I snapped, "Ask Dr. Bonnard—she referred me."

I was not only resistant to the whole business; I was stoned most of the time. One may be very glib and facile on amphetamines, and things seem to proceed with miraculous rapidity, but it all blows away, leaving no imprint.

It was utterly different at the beginning of 1966 when I myself sought out an analyst in New York, knowing that I would not survive without help. I was suspicious of Dr. Shengold at first, because he was so young. What experience of life, what knowledge, what therapeutic power, I thought, could I find in someone scarcely older than myself? I soon realized that this was someone of very exceptional caliber and character, someone who could pierce my defenses and not be deflected by my glibness, someone who felt I could tolerate and profit from intensive analysis and the intense and ambiguous feelings that transference involved.

But Shengold insisted from the start that this would only work if I gave up drugs. Drugging, he said, put me beyond the reach of analysis; he could not continue to see me unless I stopped using them. Bird might also have thought this, but he never really said it, whereas Shengold hammered it home every time I saw him. I was terrified by the idea of being "out of reach" and even more terrified that I would lose Shengold. I was still half-psychotic at times from the amphetamines I had not yet kicked. Thinking of my schizophrenic brother, Michael, I asked Shengold if I too was schizophrenic.

"No," he answered.

Was I then, I asked, "merely neurotic"?

"No," he answered.

I left it there, we left it there, and there it has been left for the last forty-nine years.

·

Nineteen sixty-six was a grim year as I struggled to give up drugs—grim too because my research was going nowhere and I was realizing that it would never get anywhere, that I did not have what it took to be a research scientist.

I would continue to seek satisfaction in drugs, I felt, unless I had satisfying—and, hopefully, creative—work. It was crucial for me to find something with meaning, and this, for me, was seeing patients.

As soon as I started clinical work in October of 1966, I felt better. I found my patients fascinating, and I *cared* for them. I started to taste my own clinical and therapeutic powers and, above all, the sense of autonomy and responsibility which I had been denied when I was still a resident in training. I had less recourse to drugs and could be more open to the analytic process.

I had one more drug high or mania in February of 1967, and this—paradoxically and unlike all my previous highs—took a creative turn and showed me what I should and could do: to write a worthwhile book on migraine and perhaps other books after this. It was not just a vague feeling of potential but a very clear, focused vision of future neurological work and writing which came to me when I was high but then stayed with me.

I never took amphetamines again—despite sometimes-intense longings for them (the brain of an addict or an alcoholic is changed for life; the possibility, the temptation, of regression never go away). And with this, I was no longer out of reach, and analysis could get somewhere.

Indeed I think it saved my life many times over. Back in 1966, my friends did not think I would make it to thirty-five, and neither did I. But with analysis, good friends, the satisfactions of clinical work and writing, and, above all, luck, I have, against all expectations, made it past eighty.

I still see Dr. Shengold twice a week, as I have been doing for almost fifty years. We maintain the proprieties—he is always "Dr. Shengold," and I am always "Dr. Sacks"—but it is because the proprieties are there that there can be such freedom of communication. And this is something I also feel with my own patients. They can tell me things, and I can ask things, which would be impermissible in ordinary social intercourse. Above all, Dr. Shengold has taught me about paying attention, listening to what lies beyond consciousness or words.

It was a huge relief to me when, in September of 1966, I stopped laboratory work and started seeing real patients at a headache clinic in the Bronx. I thought my chief concern would be headache and little else, but I soon found that the situation could be far more complex, at least in patients with so-called classical migraine, which could provoke not only intense suffering but a huge range of symptoms, almost an encyclopedia of neurology.

Many of these patients would tell me that they had seen their internists or their gynecologists or their ophthalmologists or whatever but had not received proper attention from them. This gave me a feeling of what seemed wrong with American medicine, that it consisted more and more of specialists. There were fewer and fewer primary care physicians, the base of the pyramid. My father and my two older brothers were all general practitioners, and I found myself feeling not like a super-specialist in migraine but like the general practitioner these patients should have seen to begin with. I felt it my business, my responsibility, to enquire about every aspect of their lives.

I saw one young man who had "sick headaches" every Sunday. He described the scintillating zigzags he saw before the headache, so it was easy to make a diagnosis of classical migraine. I told him we had medications for this and that if he put an ergotamine tablet under his tongue as soon as he started to see the zigzags, this might serve to abort the attack. He phoned me up in great excitement a week later. The tablet had worked, and he had no headache. He said, "God bless you, Doctor!" and I thought, "Gosh, isn't medicine easy?"

The following weekend, I did not hear from him, and, curious to see how he was getting along, I phoned him. He told me in a rather flat voice that the tablet had worked again, but he then made a curious complaint: he was bored. Every Sunday for the previous fifteen years had been devoted to migraines—his family would come, he was the center of attention—and now he missed all of that.

The week after, I got an emergency phone call from his sister saying that he was having a severe attack of asthma and was being given oxygen and adrenaline. There was a suggestion in her voice that it was perhaps my fault, that I had somehow "rocked the boat." I called on my patient later that day, and he told me that he had had attacks of asthma as a child but that these had later been "replaced" by migraine. I had missed this important part of his history by attending only to his present symptoms.

"We can give you something for the asthma," I suggested.

"No," he replied, "I'll just get something else. . . .

"Do you think I *need* to be ill on Sundays?"

I was taken aback at his words, but I said, "Let's discuss it."

We then spent two months exploring his putative need to be ill on Sunday. As we did, his migraines got less and less intrusive and finally more or less disappeared. For me, this was an example of how unconscious motives may sometimes ally themselves to physiological propensities, of how one cannot abstract an ailment or its treatment from the whole pattern, the context, the economy of someone's life.

Another patient at the headache clinic was a young mathematician who also had Sunday migraines. He would start to get nervous and irritable on Wednesday, and this would become worse by Thursday; by Friday, he could not work. On Saturday he felt tormented, and on Sunday he would have a terrible migraine. But then, towards the afternoon, the migraine would melt away. Sometimes as a migraine disappears, the person may break out in a gentle sweat or pass pints of pale urine; it is almost as if there is a catharsis at both physiological and emotional levels. As the migraine and the tension drained out of this man, he would feel himself refreshed and renewed, calm and creative, and on Sunday evening, Monday, and Tuesday he did highly original work in mathematics. Then he would start getting irritable again.

When I gave this man medication and cured him of his migraines, I also cured him of his mathematics, disrupting this strange weekly cycle of illness and misery followed by a transcendent sort of health and creativity.

No two patients with migraine were the same, and all of them were extraordinary. Working with them was my real apprenticeship in medicine.

•

The head of the migraine clinic was a man of some eminence called Arnold P. Friedman. He had written a good deal on the subject, and he had run this clinic—the first of its kind—for more than twenty years. I think Friedman took a shine to me. He thought I was bright, and I think he wanted me to be a sort of protégé. He was friendly towards me, and he arranged for me to do more clinics than everyone else and to be paid slightly more. He introduced me to his daughter, and I even wondered whether he thought of me as a potential son-in-law.

And then there came a strange episode. I would meet with him on Saturday mornings and tell him about interesting patients I had seen during the week, and on one Saturday early in 1967 I told him about a patient who did not develop a headache after the scintillating zigzag that begins many migraines but had severe abdominal pain and vomiting instead. I said that I had seen a couple of other patients like this, who had apparently switched from headache to abdominal pain, and that I wondered whether one should exhume the old Victorian term "abdominal migraine." When I said this, Friedman suddenly became a different man. He turned red and shouted, "What do you mean, talking about 'abdominal migraine'? This is a *head-*

ache clinic. The word 'migraine' comes from hemi-*crania*! It means a headache! I will not have you talk about migraines without a headache!"

I drew back, amazed. (This was one of the reasons why in the very first sentence of the book I later wrote, I emphasized that headache is never the sole symptom of a migraine—and why the second chapter of *Migraine* was entirely devoted to forms of migraine *without* a headache.) But that was a small explosion. The bigger one came in the summer of 1967.

I have described in *Hallucinations* how, in February of 1967, in an amphetamine-induced epiphany, I read Edward Liveing's 1873 book, *On Megrim*, from cover to cover and resolved to write a comparable book, a *Migraine* of my own, a *Migraine* for the 1960s, incorporating many examples from my own patients.

In the summer of 1967, after working in the migraine clinic for a year, I went back to England for a holiday, and to my own great surprise I proceeded to write a book on migraine over the course of a couple of weeks. It spilled out suddenly, without conscious planning.

I sent a telegram to Friedman from London, saying that somehow or other a book had just gushed out and that I had taken it to Faber & Faber, a British publisher (which had published a book of my mother's), and that they were interested in publishing it.[3] I hoped Friedman might like the book and write a foreword. He sent a return telegram, saying, "Stop! Hold everything."

When I came back to New York, Friedman did not look at

---

3. One reader at Faber, however, made a peculiar comment. He said, "The book is too easy to read. This will make people suspicious—professionalize it."

all friendly; he looked rather disturbed. And he almost tore the manuscript of the book out of my hands. Who did I think I was to write a book on migraine? he demanded. What presumption! I said, "I'm sorry, it just happened." He said that he would send the manuscript out for review, to someone very high up in the migraine world.

I was very taken aback by these reactions. A few days later, I saw Friedman's assistant photocopying my manuscript. I didn't pay much attention to this, but I noted it. About three weeks later, Friedman gave me a letter from the reviewer, from which all identifying characteristics of the sender had been removed. It was a letter lacking any real, constructive critical substance but full of personal and often envenomed criticism of the book's style and its writer. When I said this to Friedman, he replied, "On the contrary, he is absolutely right. This is what your book consists of; it's basically trash." He went on to say that he would not in future allow me access to any of my own notes on the patients I had seen, that everything would be locked up. He warned me not to think of going back to the book, saying that if I did, he would not only fire me but see that I never got another neurological job in America. At that time, he was chairman of the headache section of the American Neurological Association, and it would indeed have been impossible for me to get another job without his recommendation.

I mentioned Friedman's threats to my parents, hoping for their support, but my father, in what I felt was a rather cowardly way, said, "You had better not anger this man—he could ruin your life." So I suppressed my feelings for many months; these were among the worst months of my life. I continued seeing patients in the migraine clinic, and then finally, in June of 1968, I decided I could not bear it any longer. I made an arrangement with the janitor to let me into the clinic at night. Between midnight and three in the morning, I would pull out my own notes and copy what I could laboriously by hand. I then told Friedman that I wanted to take a long holiday in London, and he immediately demanded, "Are you going back to that book of yours?"

I said, "I have to."

"It'll be the last thing you do," he said.

I went back to England in a state of trepidation, literally quivering, and a week later I got a telegram from him firing me. This made the quivering worse, but then, suddenly, I had a completely different feeling. I thought, "This ape is no longer on my shoulders. I am free to do what I want."

•

Now I was free to write, but I also had an intense, literal, almost crazy feeling of an impending deadline. I was dissatisfied with my 1967 manuscript and decided to rewrite the book. It was the first of September, and I said to myself, "If I do not have the finished manuscript in Faber's hands by September 10, I shall have to kill myself." And under this threat, I started writing. Within a day or so, the feeling of threat had disappeared, and the joy of writing took over. I was no longer using drugs, but it was a time of extraordinary elation and energy. It seemed to me almost as though the book were being dictated, everything organizing itself swiftly and automatically. I would sleep for just a couple of hours a night. And a day ahead of schedule, on September 9, I took the book to Faber & Faber. Their offices were in Great Russell Street, near the British Museum, and after dropping off the manuscript, I walked over to the museum. Looking at artifacts there—pottery, sculptures, tools, and especially books and manuscripts, which had long outlived their creators—I had the feeling that I, too, had produced something. Something modest, perhaps, but with a reality and existence of its own, something that might live on after I was gone.

I have never had such a strong feeling, a feeling of having made something real and of some value, as I did with that first book, which was written in the face of such threats from Friedman and, for that matter, from myself. Returning to New York, I felt a sense of joyousness and almost blessedness. I wanted to shout, "Hallelujah!" but I was too shy. Instead, I went to concerts every night—Mozart operas and Fischer-Dieskau singing Schubert—feeling exuberant and alive.

During those excited, exalted six weeks or so in the fall of 1968, I kept on writing, thinking I might add to the migraine book a much more detailed description of the geometric patterns that may be seen in a visual aura and some speculation about what might be going on in the brain. I sent these excited addenda to William Gooddy, an English neurologist who had written a lovely foreword to the book. And Gooddy said, "No— leave it. The book is fine as it is. These are thoughts which you will return to again and again in the coming years."[4] I am glad that he protected the book against my inordinacy and exuberance, which I think had become almost manic by that point.

I worked hard with my editor to arrange the illustrations and the bibliography, and everything was ready by the spring of 1969. But when 1969 and then 1970 passed without publi-

———————

4. In fact, in 1992, I did add to the book, stimulated in part by seeing an exhibit of migraine art and in part by discussion with my friend Ralph Siegel, a very good mathematician and neuroscientist. (Twenty years after that, in 2012, I revisited the subject of migraine aura from yet another perspective when I wrote *Hallucinations*.)

cation, I felt increasingly frustrated and furious. Finally, I got a literary agent, Innes Rose, and he put some pressure on the publishers, who finally brought the book out in January of 1971 (although the imprint on the title page says 1970).

I went to London for its publication. I stayed, as always, at 37 Mapesbury, and on publication day my father came into my bedroom, pale and shaking, holding *The Times* in his hands. He said, fearfully, "You're in the papers." There was a very nice essay-review in the paper which called *Migraine* "balanced, authoritative, brilliant," or something of the sort. But so far as my father was concerned, this made no difference; I had committed a grave impropriety, if not a criminal folly, by being in the papers. In those days, one might be struck off the Medical Register in England for any indulgence in "the four *As*": alcoholism, addiction, adultery, or advertising; my father thought that a review of *Migraine* in the general press might be seen as advertising. I had gone public, made myself visible. He himself always had, or believed he had, a "low profile." He was known to and beloved by his patients, family, and friends, but not to a wider world. I had crossed a boundary, transgressed, and he feared for me. This coincided with feelings I had had myself, and in those days I often misread the word "publish" as "punish." I felt that I would be punished if I published anything, and yet I had to; this conflict almost tore me apart.

For my father, having a good name, *shem tov*, being respected by others, was all-important—more important than any particular worldly success or power. He was modest, even self-deprecatory, about himself. He downplayed the fact that he was, among other things, a remarkable diagnostician; specialists would often send him their most puzzling cases, knowing that he had an uncanny ability to come up with unexpected diagnoses.[5] But he felt secure and quietly happy in his own work, his own place in the world, the good reputation and name he bore. He hoped that all his sons, whatever we did, would also earn good names for ourselves and not dishonor the name of Sacks.

Gradually, my father, who had been so alarmed when he saw the review in *The Times*, started to be reassured when he saw good notices in the medical press, too; after all, the *British Medical Journal* and *The Lancet* had been established in the nineteenth century by doctors, for doctors. I think he started to feel, at this point, that I must have written a decent book and done the right thing persevering with it, even though it had cost me my job (and perhaps, if Friedman's power was commensurate with his threats, any other neurological jobs in America).

My mother liked the book from the start, and for the first time in many years I felt my parents were on my side, allowing that their crazy, renegade son, after years of misbehavior and folly, was now on the right clinical road—that there might be some good in me after all.

My father, who, in a joking, self-deprecating way, used to speak of himself as "the husband of the eminent gynecologist Elsie Landau" or "Abba Eban's uncle," now started to call himself "the father of Oliver Sacks."[6]

---

5. In 1972, Pop was consulted by our cousin Al Capp, who had a range of peculiar symptoms which had stymied his own doctors. My father took one look at him, as they were shaking hands, and said, "Are you on Apresoline?" (This was a drug then used to control high blood pressure.)

"Yes," Al said, surprised.

"You have SLE, systemic lupus erythematosus, caused by Apresoline," my father explained. "Fortunately this drug-induced form is entirely reversible, but if you do not stop the Apresoline, it will be fatal."

Al felt that my father, with his lightning intuition, had saved his life.

6. This went both ways, as Abba Eban wrote in an obituary for my father in *The Jewish Chronicle:*

I think that I may have underestimated my father as he underestimated himself. I was astonished and deeply moved, some years after his death, when the Chief Rabbi of England, Jonathan Sacks (no relation to our family), wrote to me:

> I knew your late father. There were times when we sat together in shul. He was a true *tzaddik*—I thought of him as one of the . . . thirty-six "hidden righteous" whose goodness sustains the world.

Even now, many years after his death, people come up to me, or write to me, speaking of my father's kindness, saying that they (or their parents or their grandparents) were patients of his during the seventy years he was in practice. Others, unsure, ask if I am related to Sammy Sacks, as everyone called him in Whitechapel. And I am happy and proud to be able to say yes.

•

When *Migraine* came out, I got a couple of rather puzzled letters from colleagues asking why I had published earlier versions of some of the chapters under the pseudonym A. P. Friedman. I wrote back, saying I had done nothing of the sort and that they should address their question to Dr. Friedman in New York. Friedman gambled foolishly on my not publishing the book,

---

I remember in 1967, after the Six-Day War, when I passed through London on my way back from the United Nations, the taxi in which I was travelling drew level with another cab at a red traffic light. My driver called out to his colleague: "Do you know who I've got in here? I've got a nephew of Dr. Sacks!"

I received the accolade without any sense of humiliation and was certainly proud for Uncle Sam. He went round telling the story for months with characteristic exuberance.

and when I did publish it, he must have realized he was in trouble. I never said a word to him, and I never saw him again.

I think Friedman had delusions of ownership, a feeling that not only did he own the whole subject of migraine but that he owned the clinic and everyone who worked there and was therefore entitled to appropriate their thoughts and their work. This painful story—painful on both sides—is not an uncommon one: an older man, a father figure, and his youthful son-in-science find their roles reversed when the son starts to outshine the father. This happened with Humphry Davy and Michael Faraday—Davy first giving every encouragement to Faraday, then trying to block his career. I am no Faraday, and Friedman was no Davy, but I think the same deadly dynamic was at work, at a much humbler level.

Helena Penina Landau, my Aunt Lennie, was born in 1892, two years before my mother. The thirteen children of my grandfather and his second wife were all close to one another and exchanged frequent letters when distance separated them, but there was a special closeness between Lennie and my mother that lasted throughout their lives.

Four of the seven sisters—Annie, Violet, Lennie, and Doogie—founded schools.[7] (My mother, Elsie, became a doctor, one of

---

7. Annie Landau, the eldest, left the comforts of London for Palestine in 1899. She knew no one in this new place but was determined to help provide a wide-ranging education for the Anglo-Jewish girls in Jerusalem, at a time when most of them were impoverished and illiterate, denied education, and pushed into teenage marriage or prostitution. They could not have found a better champion than my aunt, whose pas-

the first female surgeons in England.) Lennie had been a schoolteacher in the East End of London before she founded the Jewish Fresh Air School for Delicate Children in the 1920s. ("Delicate" could mean anything from autism to asthma or simply "nerviness.") The school was located in Delamere Forest in Cheshire, and since saying "the Fresh Air Home and School" or "JFAS" was cumbersome, we all called the school "Delamere" instead. I loved visiting it, mingling with the "delicate" children; they did not look too delicate to me. Every child (even I, a visitor) was given a square yard of ground surrounded by a low wall of stones in which we were free to plant whatever we wished. I loved botanizing with my aunt or her fellow teachers in Delamere Forest—the horsetails especially stay in my memory—and swimming in the little, shallow pond of Hatchmere ("Hatchmere of blessed memory," as my aunt once wrote, long after leaving Delamere). In the dreadful war years when I was evacuated to Braefield, I passionately longed to be at Delamere instead.

Lennie retired in 1959 after nearly forty years at Delamere, and towards the end of 1960 she found a small flat in London, but by that time I had left for Canada and the States. Four or five letters passed between us in the 1950s, but it was only when an ocean lay between us that we started to write long, frequent letters to each other.

---

sion for women's education overcame all sorts of cultural and political obstacles. Her parties, which brought together eminent Jews, Arabs, Christians, and members of the British mandate, were legendary, and the school that she directed for forty-five years left a lasting legacy on the development of modern Jerusalem. (The history of Annie Landau and her school, the Evelina de Rothschild School, is recounted in Laura S. Schor's book *The Best School in Jerusalem: Annie Landau's School for Girls, 1900–1960.*)

Lennie had sent me two letters in May of 1955—the first in response to my sending her a copy of *Seed*, a short-lived magazine (it expired after one issue) which a few friends and I had put together in my third year at Oxford.

"I am much enjoying *Seed*," Lennie wrote, "and like its whole format—the cover design, the luxurious paper, the lovely print, and the feeling for words that all you contributors have, whether grave or gay. . . . Will you be dismayed when I say how gloriously young (and of course vital) you all are."

This letter, like all her letters, opened with "Darling Bol" (occasionally "Boliver"), whereas my parents, more soberly, would write "Dear Oliver." I did not feel she used the word "Darling" lightly; I felt very loved by her, and I loved her intensely too, and this was a love without ambivalence, without conditionality. Nothing I could say could repel or shock her; there seemed no limit to her powers of sympathy and understanding, the generosity and spaciousness of her heart.

She would send me postcards when she traveled. "Here I am basking in brilliant sunshine in Grieg's garden," she wrote in 1958, "looking down at a magical fjord. No wonder he was inspired to make music. (What a pity you're not here. There are a number of pleasant young men in the party . . . we're a quite civilized bunch of mixed ages and sexes.)"

Coincidentally, I too went to Norway in 1958 and stayed on a tiny island called Krokholmen in Oslo fjord (where a friend of mine, Gene Sharp, had a little house). "When I got your idyllic card from Krokholmen," Lennie wrote, "I wished I could have come and been Man Friday to your Robinson Crusoe." She ended her letter by wishing me "all that's splendid for your finals in December."

•

Nineteen sixty was a year of profound change for both of us. Lennie left Delamere after heading the school for nearly forty years, and I left England. I was twenty-seven and she was sixty-seven, but both of us felt we were embarking on a new life. Lennie decided on a leisurely trip around the world before settling in London, and I was already in Canada when I got her letter from her ship, the *Strathmore*.

"We get to Singapore tomorrow," she wrote. "For a couple of days [after leaving Perth], besides disporting dolphins, we had magnificent albatrosses following us . . . wonderfully graceful, dipping and rising in their flight, with a tremendous wingspan."

In October, when I had started working in San Francisco, she wrote, "I was delighted to get your letter . . . you certainly seem to have found a more satisfying outlet for your restless and searching spirit. . . . I do miss you." Conveying a message from my mother, she added, "Her favorite indoor sport is still packing parcels for you!"

In February of 1961, Lennie wrote of a recurrent problem with my brother Michael: "I have never seen Michael as alarming as at this time, and to my self-disgust, my pity changed to revulsion and fear, and your mother's fierce protection somehow suggested (although I hoped my feelings didn't come through) that everybody was out of step except Michael."

Lennie had been very fond of Michael when he was a boy; like Auntie Annie, she admired his precocious intellect and would bring him whatever books he desired. But now my parents, she felt, were denying the gravity—and danger—of the situation. "In the last weeks before he went back to Barnet [a psychiatric hospital] I feared for their lives. What a pathetic, blighted life." He was thirty-two.

After much searching—rents were high in London, and Lennie had never been able to save much ("Like you, money slips through my fingers")—Lennie had found a place in Wembley: "I think you'll like this little flat of mine. I like having my own home, and now am partly compensated for having lost Delamere. As I write, the almond trees outside my window are in bloom, there are crocuses, snowdrops and some early daffodils, and even a chaffinch pretending that Spring is here."

Going to plays was much easier now that she was in London, she wrote. "Looking forward to going to Harold Pinter's *The Caretaker* tomorrow evening. . . . These new young writers haven't the polished and rounded phrases of my generation, but they've got something real to say, and they say it with vigour." She was enjoying, too, the rising generation of great-nephews and great-nieces, as she had with my generation—especially my brother David's children.

•

In May of 1961, I sent her the manuscript of "Canada: Pause, 1960," drawing on my travels across Canada, and another journal ("99") about a night ride from San Francisco to Los Angeles. These were in a sense my first "pieces"—self-conscious and precious in tone, but pieces which I hoped might be published one day.

"I received your amazing excerpts from your journals," Len wrote. "I found the whole thing breathtaking. I was suddenly conscious that I was gasping physically." I had not shown these pieces to anyone else but Thom Gunn, and Auntie Len's enthusiasm, not unmixed with criticism, was crucially important to me.

Lennie was especially fond of Jonathan Miller and his wife, Rachel, as they were of her. Jonathan, she wrote, "remains the same unspoilt, simple, complex, brilliant, lovable, untidy genius—like you. . . . We had a long jaw together one afternoon when we were both at Mapesbury. . . . How he gets into his one life all that he does is incredible."

She enjoyed the photographs of California I sent. Riding far afield on my motorbike, always with a camera, I sent her photographs of California landscapes. "What lovely pictures," she wrote. "So extraordinarily like the Grecian scenery I saw during my tantalizing brief visit on my way home from Australia. . . . Be careful on that steed of yours!"

Len liked "Travel Happy" when I sent it to her early in 1962 but thought I was too free with the truckers' "fucks" and "shits." I found these exotic, very American—in England, we never went beyond "bugger"—but Len thought them "boring when written so frequently."

In November of 1962, she wrote, "Your mum has begun to operate again [she had fractured a hip earlier in the year] which delights her, and she no longer feels frustrated. Your Pop is his same lovable, crazy, untidy self, leaving little bits of kindness in the shape of specs, syringes, notebooks and so on wherever he goes. And eager and willing hands gather them up and deliver them, as if it were the greatest honour in the world."

Lennie was thrilled that I was presenting a paper at a neurology meeting—my first sally into academia—but "*not* thrilled that you're working up to an enormous weight again—you're such a good-looking bloke when you're normal."

I mentioned to her, a couple of months later, that I had been in a depression. "I know we all suffer them at times," Len wrote. "Well, don't have any more. You've got so much in your favour—brains, charm, presentability, a sense of the ridiculous, and a whole gaggle of us who believe in you."

Len's belief in me had been important since my earliest

years, since my parents, I thought, did *not* believe in me, and I had only a fragile belief in myself.

Emerging from my depression, I sent Lennie a parcel of books, and while reproaching me for my "extravagance," she replied, "Here are all my thanks to my favorite nephew." (I liked the sound of that, for Lennie was certainly my favorite aunt.) She continued, "Picture me cozily by the fire, a bowl of Cox's Orange Pippins to hand, immersed in Henry James's elegant richness, and then suddenly realizing that the small hours have set in." This letter was illegible in parts—"No, my writing hasn't become senile, I've been trying to break in a new fountain pen, having lost my precious fifty-year-old one."

She always wrote with a broad-nibbed fountain pen (as I still do, fifty years later). "Darling Bol," she ended, "may you be happy."

·

"I've heard of your battle with the waves, you crazy loon," she wrote in 1964. I had written to her of how I dislocated a shoulder when flung on Venice Beach by a giant wave and how my friend Chet had pulled me out.

She hoped I would send her some of my papers on neurology, "of which I'll understand not one word, but will glow with loving pride at my ridiculous, brilliant and altogether delightful nephew."

And so our letters continued, seven or eight a year. I wrote to Lennie of leaving California and my first impressions of New York:

This is really a marvelous city, rich, exciting, unlimited in range and depth—as London is; although the two cities are profoundly different. New York is punctate, scintillating, the way all cities look from a plane at night: it is a *mosaic* of qualities and people and dates and styles, a sort of enormous urban jigsaw. Whereas London has much the quality of an *evolved* city, the present like a transparency overlying the wafers of the past, layer on layer, extended in time, like Schliemann's Troy, or the crust of the earth. But then again, for all its sparkling synthetic quality, New York is strangely old-fashioned, archaic. The huge girders of the "El" are a railway-phantasy of the 1880's, the crayfish tail of the Chrysler building a pure Edwardian vanity. I can't see the Empire State Building without the vast silhouette of King Kong scrambling up its sides. The East Bronx is like Whitechapel in the early Twenties (before the diaspora to Golders Green).

Len wrote of family events, of books she had read and plays she had seen, and especially of her robust walking trips. She remained an enthusiastic hill walker into her seventies and now had the leisure to explore the wilder parts of Ireland, Scotland, and Wales.

Along with her letters came parcels of Blue Vinny, a blue cheese from a single dairy farm in Dorset; I adored this and rated it above Stilton. I loved the slightly smelly parcels that arrived every month, each containing a quarter of a Blue Vinny wheel. She had started sending me these in my Oxford days, was still doing so fifteen years later.

In 1966, Len wrote to me about my mother's second hip operation. She wrote, "Your mum has had a rough week. . . . Your pop was a very worried man." But all went well—my mother moved to crutches, then a stick—and the following month Lennie wrote, "Her grit and determination are incredible." (*All* the Landau siblings, it seemed to me, were strongly endowed with grit and determination.)

I wrote to Lennie early in 1967, after I had read Liveing's book *On Megrim* and determined to write my own book on the subject. Lennie was excited by this; she had felt, since my boyhood days, that I could and *should* be "a writer." I wrote to her of Friedman's reaction to my manuscript and how my father thought that I would do well to defer to him, but Lennie, with her characteristic Landau clarity and tough-mindedness, did not agree.

"Your Dr. Friedman," she wrote in October of 1967, "sounds a most unpleasant piece of work, but don't let him get under your skin. Keep hold of your faith in yourself."

•

In the autumn of 1967, my parents stopped in New York; they were on the way back from Australia, where they visited my eldest brother, Marcus, and his family. Our parents had worried about me, one way and another, and now they could see for themselves that I was enjoying professional life, appreciating and being appreciated by my patients—my brother David had visited New York a few months earlier and reported that I was "adored" by my patients—and writing about the extraordinary postencephalitic patients I was now seeing in New York. A few weeks later, Lennie wrote, "Your mum and pop came home completely refreshed at having seen their youngest and eldest in their respective habitats," adding that Marcus, in Australia, had written her a letter of "lyrical ecstasy" about his baby daughter.

By 1968, larger threats were looming—the Vietnam War and an intensification of the draft; I was summoned for a military interview but managed to persuade the powers that I was not suitable material for the army.

"Let me tell you how relieved we all are that you're to remain a civilian," Len wrote. "This Vietnamese War grows more hor-rific daily, and the web becomes more and more entangled. . . . What are your views on the horrible muddle the world over (as well as the occasional good happenings)? Do write and let me know how you are."

Awakenings

In the fall of 1966, I started seeing patients at Beth Abraham, a chronic disease hospital affiliated with the Albert Einstein College of Medicine. I soon realized that among its five hundred residents, some eighty patients, dispersed in various wards, were survivors of the extraordinary encephalitis lethargica (or sleepy sickness) pandemic which had swept the world in the early 1920s. The sleepy sickness had killed many thousands outright, and those who had seemingly recovered often came down, sometimes decades later, with strange postencephalitic syndromes. Many were frozen in deeply parkinsonian states, some stuck in catatonic postures—not unconscious but with their consciousness suspended at the point where the disease had closed in on certain parts of the brain. I was amazed when I heard that some of the patients had been like this for thirty or forty years—indeed, that the hospital had originally been opened in 1920 for these first victims of the encephalitis lethargica.

During the 1920s and 1930s, hospitals had been built or converted around the world to accommodate postencephalitic patients; one such, the Highlands Hospital in north London, had originally been a fever hospital with dozens of pavilions spread out over many acres but was then used to house nearly twenty thousand postencephalitics. But by the late 1930s, most of those affected had died, and the disease itself—once front-page news—was all but forgotten. There were very few reports in the medical literature about the strange postencephalitic syndromes that might not become apparent until decades later.

The nurses, who knew these patients well, were convinced that behind their statuesque appearance—locked in, imprisoned—there were intact minds and personalities. The nurses also mentioned that the patients might have occasional, very brief liberations from their frozen states; music, for example, might animate the patients and allow them to dance, even though they could not walk, or to sing, even though they could not speak. On rare occasions, moreover, some might move spontaneously, suddenly, and with lightning speed, so-called *kinesia paradoxa.*

What fascinated me was the spectacle of a disease that was never the same in two patients, a disease that could take any possible form—one rightly called a "phantasmagoria" by those who had studied it in the 1920s and 1930s. It was a syndrome that included an enormous range of disturbances occurring at every level of the nervous system, a disorder that could show far better than any other how the nervous system was organized, how brain and behavior worked at their more primitive levels.

When I wandered among my postencephalitic patients, I sometimes felt like a naturalist in a tropical jungle, sometimes, indeed, in an *ancient* jungle, witnessing prehistoric, pre-human behaviors—grooming, clawing, lapping, sucking, panting, and a whole repertoire of strange respiratory and phonatory behaviors. These were "fossil behaviors," Darwinian vestiges of earlier times brought out of physiological limbo by the stimulation of primitive brain-stem systems, damaged and sensitized by the encephalitis in the first place, and now "awakened" by L-dopa.[1]

I spent a year and a half observing and taking notes, sometimes filming the patients and recording them, and in that time I got to know them not only as patients but as people. Many of them had been abandoned by their families and had no contact with anyone but the nursing staff. It was not until I dug out their charts from the 1920s and 1930s that I could confirm their diagnoses, and at this point I asked the director of the hospital if we could move some of them together into a single ward, in the hope that this would allow a community to form.

I felt from the start that I was seeing individuals in an unprecedented state and situation, one which had never been described, and within weeks of encountering them in 1966, I pondered writing a book about them; I thought of adopting one of Jack London's titles, *The People of the Abyss*. This sense of the dynamics of illness and life, of the organism or subject striving to survive, sometimes under the strangest and darkest circumstances, was not a viewpoint which had been emphasized when I was a student or resident, nor was it one I found in the current medical literature. But when I saw these postencephalitic patients, it was clearly and overwhelmingly true. What had been dismissed disparagingly by most of my colleagues ("chronic hospitals—you'll never see anything interesting in

1. Macdonald Critchley, in his biography of William R. Gowers, the Victorian neurologist (and amateur botanist), writes, "To him the neurological sick were like the flora of a tropical jungle." Like Gowers, I sometimes see my patients with unusual disorders as different and extraordinary beings, forms of life.

those places") revealed itself as the complete opposite: an ideal
situation for seeing entire lives unfold.

•

It had been established in the late 1950s that the parkinsonian
brain was deficient in the transmitter dopamine and that it
might therefore be "normalized" by raising the level of dopamine. But attempts to do this by giving L-dopa (a precursor of
dopamine) in milligram quantities had unclear effects, and it
was not until George Cotzias, with great audacity, gave doses a
thousand times greater to a group of patients with Parkinson's
disease that extraordinary therapeutic effects were seen. With
the publication of his results in February 1967, the outlook
for patients with Parkinson's disease was changed at a stroke:
patients hitherto able to look forward only to miserable and
increasing disability might be transformed by the new drug.
The atmosphere was electric with excitement, and I wondered
whether L-dopa could help my own, very different patients.

Should I give L-dopa to our patients at Beth Abraham? I hesitated; they did not have ordinary Parkinson's disease but a postencephalitic disorder of far greater complexity, severity, and
strangeness. How would *these* patients, with their so-different
disease, react? I felt I had to be cautious—almost exaggeratedly
so. Might L-dopa activate neurological problems some of these
patients had had in the early years of their illness, before they
were encased in parkinsonism?

In 1967, with some trepidation, I applied to the Drug Enforcement Administration for a special investigator's license to use
L-dopa, still an experimental drug at the time. The license took
several months to come, and for various reasons it was not
until March of 1969 that I embarked on a ninety-day double-
blind trial with six patients. Half of them would receive a pla-

cebo, but neither they nor I knew who would receive the real drug.

But within a few weeks, the effects of L-dopa were clear and spectacular. I could infer from the precise 50 percent failure rate there was no significant placebo effect whatever. I could no longer, in good conscience, continue the placebo but decided to make L-dopa available for any patient ready to try it.[2]

At first, nearly all the patients' responses were happy ones; there was an astonishing, festive "awakening" that summer as they burst into explosive life after having been almost inanimate for decades.

But then almost all of them ran into trouble, developing not only specific "side effects" of L-dopa but certain general patterns of trouble too: sudden and unpredictable fluctuations of response and extreme sensitivity to L-dopa. Some of the patients would react differently to the drug each time we tried it. I tried altering the doses, titrating them carefully, but this no longer worked; the "system" now seemed to have a dynamic of its own. There seemed to be, with many of the patients, nothing between too much L-dopa and too little.

I thought of Michael and his problems with tranquilizers (which damped down dopamine systems, while L-dopa acti-

---

2. Around this time, I had a discussion with my chief at Einstein, Labe Scheinberg. "How many patients do you have on L-dopa?" he asked me.

"Three, sir," I replied eagerly.

"Gee, Oliver," Labe said, "I have three *hundred* patients on L-dopa."

"Yes, but I learn a hundred times as much about each patient as you do," I replied, stung by his sarcasm.

Series are needed—all sorts of generalizations are made possible by dealing with populations—but one needs the concrete, the particular, the personal too, and it is impossible to convey the nature and impact of any neurological condition *without* entering and describing the lives of individual patients. *entirety of life, not just drug's affect on their brain.*

vated them) when I was trying to titrate my own patients, finding the incorrigible limitations of any purely medical or medicational approach when dealing with brain systems which had seemingly lost their usual resilience or latitude.[3]

•

When I was a resident at UCLA, neurology and psychiatry were presented as almost unrelated disciplines, but when I emerged from residency to encounter the full reality of patients, I often found I had to be as much a psychiatrist as a neurologist. I had felt this strongly with my migraine patients, and I encountered it overwhelmingly with the postencephalitics, for they had a myriad of disorders both "neurologic" and "psychiatric": parkinsonism, myoclonus, chorea, tics, strange compulsions, urges, obsessions, sudden "crises," and gusts of passion. A purely neurological or a purely psychiatric approach with such patients would lead nowhere; the neurological and the psychiatric had to be conjoined.

The postencephalitics had been in a state of suspension for decades—suspension of memory, perception, and consciousness. They were coming back to life, to full consciousness and mobility. Would they find themselves, like Rip Van Winkle, anachronisms in a world that had moved on?

When I gave L-dopa to these patients, their "awakenings" were not only physical but intellectual, perceptual, and emo-

3. In August of 1969, the "awakenings" of my postencephalitic patients hit *The New York Times* in the form of a long, illustrated article by Israel Shenker. He described what I called a "yo-yo effect" in some of my patients—sudden oscillations of drug effects—a phenomenon not described by other colleagues, or in other patients, until several years later (and then called an "on-off effect"). While L-dopa was presented as "a miracle drug," I commented in the article how crucial it was to pay attention to the entirety of patients' lives and situations and not just the effects of a drug on their brains.

tional too. Such a global awakening or animation was in contradiction to concepts of neuroanatomy in the 1960s, a neuroanatomy that saw the motor, the intellectual, and the affective in quite separate and noncommunicating compartments of the brain. The anatomist in me, subservient to this notion, said, "This can't be. Such an 'awakening' should not happen."

But clearly it *was* happening.

The Drug Enforcement Administration wanted me to fill out standardized inventories of symptoms and responses to the drug, but what was going on was so complex in both neurological and human terms that such inventories could not begin to accommodate the reality of what I was witnessing. I felt a need to keep detailed notes and journals, as did some of the patients. I started carrying a tape recorder and a camera, and later a little Super 8 movie camera, because I knew that what I was seeing might never be seen again; it was crucial to have a visual record.

Some of the patients would sleep for much of the day but be wide awake at night, and this meant that I too had to have a twenty-four-hour schedule. Although this led to sleep deprivation, it gave me a feeling of closeness to them, and it also allowed me to be on night call for *all* of the five hundred patients at Beth Abraham. This was a job that might involve treating a patient with acute heart failure, sending another to an emergency room, or requesting an autopsy if a patient had died. While there was normally a different on-call doctor every night, I thought I might as well be on permanent night call, so I volunteered for this.

The administrators at Beth Abraham liked the idea and offered me, at a very nominal rate, an apartment in a house next door to the hospital—the apartment normally reserved for whichever physician was on call. This worked well for everyone: most of the other physicians hated being on call, and I was delighted to have an apartment which was always open to my patients. Staff members—psychologists, social workers, physiotherapists, speech therapists, music therapists, among them—dropped by frequently to discuss the patients. Almost every day, there were fertile, exciting discussions about the unprecedented events unfolding before us, which demanded unprecedented approaches from us all.

James Purdon Martin, an eminent neurologist in London who had decided to spend his retirement observing and working with the postencephalitic patients at the Highlands Hospital, had published a remarkable book on their balance and postural abnormalities in 1967. He made a special journey to New York in September of 1969 to see my patients; it could not have been easy for him, because he was in his seventies by then. He was fascinated by seeing the patients on L-dopa and said he had seen nothing like it since the acute days of the epidemic fifty years earlier. "You must write about all of this, in great detail," he insisted.

In 1970, I started to write about the postencephalitics in what for me has always been a favorite form, the letter to the editor. In one week, I sent four letters to the editor of *The Lancet*, and these were immediately accepted for publication. But my boss, the medical director at Beth Abraham, was not pleased. He said, "Why are you publishing these things in England? You're here in America; you must write something for *The Journal of the American Medical Association*. Not these letters about individuals, but a statistical survey of all the patients and how they are doing."

In the summer of 1970, then, in a letter to *JAMA*, I reported my findings, describing the total effects of L-dopa in sixty patients whom I had maintained on it for a year. Nearly all of them, I noted, had done well at first, but almost all of them sooner or later had escaped from control, had entered complex, sometimes bizarre, and unpredictable states. These could not, I wrote, be seen as "side effects" but had to be regarded as integral parts of an evolving whole.

JAMA published my letter, but while I had got plenty of positive responses from colleagues in reply to my letters in *The Lancet*, my letter in *JAMA* was greeted by a strange, rather frightening silence.

The silence was broken a few months later, when the entire letters section in one October issue of *JAMA* was devoted to highly critical and sometimes angry responses from various colleagues. Basically, they said, "Sacks is off his head. We ourselves have seen dozens of patients, but we've never seen anything like this." One of my colleagues in New York said that he had seen more than a hundred parkinsonian patients on L-dopa but had never seen any of the complex reactions I described. I wrote back to him saying, "Dear Dr. M., Fifteen of your patients are now under my care at Beth Abraham. Would you like to visit them and see how they are doing?" I did not get a reply.

It seemed to me then that some of my colleagues were downplaying some of the negative effects of L-dopa. One letter said that even if what I described was real, I should not publish it, because it would "negatively impact the atmosphere of optimism necessary for a therapeutic reaction to L-dopa."

I thought it was improper of *JAMA* to publish these attacks without giving me an opportunity to respond to them in the same issue. What I would have made clear was the extreme sensitivity of postencephalitic patients, a sensitivity which caused them to react to L-dopa much sooner and more dramatically than patients with ordinary Parkinson's disease. Thus I saw effects in my patients within days or weeks which my colleagues treating ordinary Parkinson's disease would not see for years.

But there were deeper issues, too. In my letter to *JAMA*, not only had I cast doubt on what had appeared at first to be the extremely simple matter of giving a drug and being in control of its effects, I had cast doubt on predictability itself. I had cast contingency as an essential, unavoidable phenomenon that emerged with the continuing administration of L-dopa.

I knew that I had been given the rarest of opportunities; I knew that I had something important to say, but I saw no way of saying it, of being faithful to my experiences, without forfeiting medical "publishability" or acceptance among my colleagues. I felt this most keenly when a long paper I had written about the postencephalitics and their responses to L-dopa was rejected by *Brain*, the oldest and most respected journal of neurology.

In 1958, while I was a medical student, the great Soviet neuropsychologist A. R. Luria had come to London to give a talk about speech development in a pair of identical twins, and he combined powers of observation, theoretical depth, and human warmth in a way which I thought revelatory.

In 1966, after arriving in New York, I read two of Luria's books, *Higher Cortical Functions in Man* and *Human Brain and Psychological Processes*. The latter, which contained very full case histories of patients with frontal lobe damage, filled me with admiration.[4]

In 1968, I read Luria's *Mind of a Mnemonist*. I read the first thirty pages thinking it was a novel. But then I realized that it was in fact a case history—the deepest and most detailed case history I had ever read, a case history with the dramatic power, the feeling, and the structure of a novel.

Luria had achieved international renown as the founder of neuropsychology. But he believed his richly human case histories were no less important than his great neuropsychological treatises. Luria's endeavor—to combine the classical and the romantic, science and storytelling—became my own, and his "little book," as he always called it (*The Mind of a Mnemonist* is only a hundred and sixty small pages), altered the focus and direction of my life, by serving as an exemplar not only for *Awakenings* but for everything else I was to write.

In the summer of 1969, having worked for eighteen hours a day with the postencephalitics, I took off for London in a state of exhaustion and excitement. Inspired by Luria's "little book," I spent six weeks in my parents' house, where I wrote the first nine case histories of *Awakenings*. When I offered them to my publishers at Faber & Faber, they said they were not interested.

I also wrote a forty-thousand-word manuscript on postencephalitic tics and behaviors and, additionally, planned a treatise titled "Subcortical Functions in Man," a complement to

---

4. And fear, for as I read it, I thought, what place is there for me in the world? Luria has already seen, said, written, and thought anything I can ever say, or write, or think. I was so upset that I tore the book in two (I had to buy a new copy for the library, as well as a copy for myself).

Luria's *Higher Cortical Functions in Man.* These, too, were turned down by Faber.

When I first came to Beth Abraham in 1966, it housed, in addition to the eighty-odd postencephalitic patients, hundreds of patients with other neurological diseases— younger patients with motor neuron disease (ALS), syringomyelia, Charcot-Marie-Tooth disease, etc.; older patients with Parkinson's disease, strokes, brain tumors, or senile dementia (in those days, the term "Alzheimer's disease" was reserved for rare patients with a pre-senile dementia).

The head of neurology at Einstein asked me to use this unique population of patients to introduce its medical students to neurology. I would have eight or nine medical students at a time—students who had a special interest in neurology and who could come on Friday afternoons for two months (there were also sessions on other days to accommodate Orthodox students who could not come on Fridays). The students would learn not only about neurological disorders but what it meant to be institutionalized and to live with a chronic disability. We would ascend, week by week, from disorders of the peripheral nervous system and spinal cord to the brain stem and cerebellar disorders and then to movement disorders and finally to disorders of perception, language, thought, and judgment.

We always started with bedside teaching, gathering around a patient's bed to elicit his history, ask him questions, and examine him. I would stand by the patient, not intruding for the most part, but making sure the patient was always treated with respect, courtesy, and full attention.

The only patients I introduced the students to were ones I

knew well and who had agreed to being questioned and examined by the students. Some of them were born teachers themselves. Goldie Kaplan, for one, had a rare congenital condition affecting the spinal cord, and she would say to students, "Don't try to memorize 'syringomyelia' from your textbooks—think of *me*. Observe this large burn on my left arm, where I leaned against a radiator without feeling heat or pain. Remember the twisted way I sit in a chair, the difficulty I have with speech because the syrinx is starting to reach into my brain stem. I *exemplify* syringomyelia!" she would say. "Remember *me*!" All the students did, and some, writing to me many years later, would mention Goldie, saying they could still see her in their mind's eye.

After three hours of seeing patients, we would break for tea served in my crowded little office, its walls covered by a palimpsest of papers I had pinned up: articles, notes and thoughts, poster-size diagrams. Then, weather permitting, we might go across the road to the New York Botanical Garden, where we would settle under a tree and talk about philosophy and life generally. We got to know one another well over the course of our nine Friday afternoons.

At one point, the neurology department asked me to test and grade my students. I submitted the requisite form, giving all of them A's. My chairman was indignant. "How can they all be A's?" he asked. "Is this some kind of a joke?"

I said, no, it wasn't a joke, but that the more I got to know each student, the more he seemed to me distinctive. My A was not some attempt to affirm a spurious equality but rather an acknowledgment of the uniqueness of each student. I felt that a student could not be reduced to a number or a test, any more than a patient could. How could I judge students without seeing them in a variety of situations, how they stood on the ungradable qualities of empathy, concern, responsibility, judgment?

Eventually, I was no longer asked to grade my students.

Occasionally, I would have a single medical student for longer periods. One of them, Jonathan Kurtis, visited me recently and told me that now, more than forty years later, the only thing he remembered from his medical student days was that three-month period he spent with me. I would sometimes tell him to see a patient with, say, multiple sclerosis—to go to her room and spend a couple of hours with her. Then he had to give me the fullest possible report not only on her neurological problems and ways of living with them but on her personality, her interests, her family, her entire life history.[5]

We would discuss the patient and the "condition" in more general terms, and then I would suggest further reading; Jonathan was struck by the fact that I would often recommend original (often nineteenth-century) accounts. No one else in medical school, Jonathan said, ever suggested that he read such accounts; they were dismissed, if mentioned at all, as "old stuff," obsolete, irrelevant, of no use or interest to anyone but a historian.

•

Nursing aides, escorts, orderlies, and nurses at Beth Abraham (as in hospitals everywhere) worked long hours and were poorly paid, and in 1972 their union, local 1199, called a strike. Some of the staff had worked at the hospital for many years and had

---

5. Perhaps I was influenced here by something William James had written of his own teacher Louis Agassiz—how Agassiz "used to lock a student up in a room full of turtle shells, or lobster shells, or oyster shells, without a book or work to help him, and not let him out till he had discovered all the truths which the objects contained."

become very attached to their patients. I spoke to a number of them as they stood on the picket line, and they told me how conflicted they felt at abandoning their patients; some were weeping.

I feared for some of the patients, especially those who were immobile and needed frequent turning in bed to prevent bedsores, as well as passive range-of-movement exercises for their joints, which would otherwise freeze up. A single day without being turned and "ranged" could start these patients on a downhill course, and it looked as if the strike might last a week or more.

I phoned a couple of my students, explained the situation, and asked if they could help. They agreed to convene a meeting of the student board to discuss the matter. They called back two hours later, saying apologetically that the student board could not condone strikebreaking as a group. But, they added, individual students could follow their own consciences; the two students I had called said they would be right over.

I walked with them through the picket line—the striking workers allowed us through—and we spent the next four hours turning patients, ranging their joints, and taking care of their toilet needs, at which point the two students were relieved by another pair of students. It was backbreaking, round-the-clock work, and it made us realize how hard the nurses and aides and orderlies worked in their normal routines, but we managed to prevent skin breakdown or any other problems among the more than five hundred patients.

Work and wage issues were finally resolved, and the staff came back to work ten days later. But that last evening when I walked to my car, I found the windscreen had been smashed. Attached to it was a large, handwritten notice that said, "We love you, Dr. Sacks. But you have been a strike-breaker." They had waited, though, until the end of the strike, to allow me and the students to care for the patients.[6]

As one grows older, the years seem to blur into one another, but 1972 remains sharply etched in my memory. The previous three years had been a time of overwhelming intensity, with the awakenings and tribulations of my patients; such an experience is not given to one twice in a lifetime, nor, usually, even once. Its preciousness and depth, its intensity and range, made me feel I had to articulate it somehow, but I could not imagine an appropriate form, one which could combine the objectivity of science with the intense sense of fellow feeling, the closeness I had with my patients, and the sheer wonder (and sometimes tragedy) of it all. I entered 1972 with a sharp sense of frustration, an uncertainty as to whether I would ever find a way of binding the experience and forging it into some organic unity and form.

I still regarded England as my home and my twelve years in the States as little more than a prolonged visit. It seemed to me that I needed to return, to go home to write. "Home" meant many things: London; the large, rambling house in Mapesbury Road where I was born and where my parents, now in their seventies, still lived with Michael; and Hampstead Heath, where I used to play as a child.

---

6. This would not be the case in a 1984 strike, in which no one was allowed to cross the picket line for forty-seven days. Many patients suffered; thirty of them, I wrote in a letter to my father, died from neglect during that time, even though temporary employees and administrators came in to care for them.

I decided to take the summer off and to get myself a flat on the edge of Hampstead Heath, within easy range of the walks, the mushroomed woods, the swimming ponds I loved, and equally easy range of Mapesbury Road. My parents would be celebrating their golden wedding anniversary in June, and there would be a gathering of the families—not only my three brothers and myself, but my parents' siblings, nieces, nephews, and the far-flung cousinhood.

But I had a more specific reason for being close: my mother was a natural storyteller. She would tell medical stories to her colleagues, her students, her patients, her friends. And she had told us—my three brothers and me—medical stories from our earliest days, stories sometimes grim and terrifying but always evocative of the personal qualities, the special value and valor, of the patient. My father, too, was a grand medical storyteller, and my parents' sense of wonder at the vagaries of life, their combination of a clinical and a narrative cast of mind, were transmitted with great force to all of us. My own impulse to write—not to write fiction or poems, but to chronicle and describe—seems to have come directly from them.

My mother had been fascinated when I told her about my postencephalitic patients and their awakenings and tribulations when I put them on L-dopa. She had been urging me to write their stories, and in the summer of 1972 she said, "Now! This is the time."

I spent each morning walking and swimming on the Heath, and each afternoon writing or dictating the stories of *Awakenings*. Every evening I would stroll down Frognal to Mill Lane and then to 37 Mapesbury Road, where I would read the latest installment to my mother. She had read to me by the hour when I was a child—I first experienced Dickens, Trollope,

D. H. Lawrence, through her reading—and now she wanted me to read to her, to give full narrative form to the stories she had already heard in bits and pieces. She would listen intently, always with emotion, but equally with a sharp critical judgment, one honed by her own sense of what was clinically real. She tolerated, with mixed feelings, my meanderings and ponderings, but "ringing true" was her ultimate value. "That doesn't ring true!" she would sometimes say, but then, more and more, "Now you have it. Now it rings true."

In a sort of way, then, we wrote the case histories of *Awakenings* together that summer, and there was a sense of time arrested, of enchantment, a privileged time-out from the rush of daily life, a special time consecrated to creation.

My flat in Hampstead Heath was also in easy walking distance of Colin Haycraft's office in Gloucester Crescent. I remember seeing Colin in 1951, when I was a freshman at Queen's. He was in his final year—a short, energetic figure in his scholar's gown, already Gibbonian in confidence and mannerism, but agile and darting in movement and said to be a brilliant racquets-player as well as a classical scholar. But we did not actually meet until twenty years later.

I had written the first nine case histories of *Awakenings* in the summer of 1969, but these had been turned down by Faber & Faber, a rejection which threw me and made me wonder if I would ever complete or publish a book again.[7] I had put this manuscript away and subsequently lost it.

---

7. Raymond Greene, of Heinemann (who had reviewed *Migraine* warmly when it came out in early 1971), wanted to commission me to write a book on parkinsonism "just like" *Migraine*. This both encouraged and discouraged me, because I did not want to repeat myself; I felt a quite different sort of book was called for, but I had no idea *what* sort of book it should be.

Colin Haycraft, by this time, had a much respected publishing firm, Duckworth, just across the road from Jonathan Miller's house. In late 1971, seeing my quandary, Jonathan had taken Colin a carbon copy of the first nine case histories; I had completely forgotten that he had a copy.

Colin liked the case histories and urged me to write more. This excited me but scared me too. Colin pressed, gently; I demurred; he drew back, waited, moved in once again; he was very sensitive, very delicate with my diffidence and anxieties. I prevaricated for six months.

Sensing that I needed a further push, Colin, in the impulsive-intuitive way he often did things, put the typescript Jonathan had given him into proofs. He did this without warning, without consulting me, in July. It was a most generous, not to say extravagant, act—what guarantee had he that I would ever continue my writing?—and also a crucial act of faith. This was before the advent of digital typesetting, and he had gone to considerable expense to produce these long galley proofs. And that was proof for me that he really thought the book was good.

I secured a shorthand typist; I had a whiplash injury at the time, having rushed up the cellar stairs and cracked my head on a low beam, which caused such wasting of the right hand that I could not hold a pen. I forced myself to work and dictate daily—a duty which rapidly became a delight as I got into the work more and more. Dictation is not quite the right word. I settled myself on the couch, wearing my cervical collar, looked through my notes, and then told my stories to the typist, watching her facial expressions closely as she transcribed them into shorthand. Her reactions were crucial: I was talking not into a machine but to *her*; the scene was rather like Scheherazade in reverse. Every morning, she would bring me the previous day's transcripts, beautifully typed, and I would read these to my mother in the evening.

And almost every day, I sent completed bunches of typescript down to Colin, which we would then go over in minute detail. We spent hours that summer closeted together. And yet I see, from letters between us, that we still preserved a considerable formality: he was always "Mr. Haycraft"; I was always "Dr. Sacks." On August 30, 1972, I wrote:

> Dear Mr. Haycraft,
>
> I enclose, herewith, five more case histories. The sixteen histories thus far come to about 240 pages in aggregate, which would be between 50 and 60,000 words. . . . I am thinking of adding four further ones . . . but I will defer here, of course, to your judgment in the matter. . . .
>
> I have tried to move from piles and compilations of medical lists to stories, but obviously without complete success. You're so right about the shape of Art and the shapelessness of Life—perhaps I should have had a keener, cleaner line or theme in them all, but they are so complex, like tapestries. To some extent these are crude ore, which others (including myself) can dig in and refine later.
>
> With kindest regards,
> Oliver Sacks

A week later, I wrote:

> Dear Mr. Haycraft,
>
> I have spent several days on an introduction . . . which I herewith enclose. I only seem to find the right way after making every possible blunder, and finally exhausting all the wrong

ways. . . . I need to talk to you again soon . . . as always, because
you help me to unmuddle.

In the summer of 1972, Mary-Kay Wilmers, a neighbor of
Colin's in Gloucester Crescent and the editor of *The Listener,*
a weekly paper published by the BBC, invited me to write an
article about my patients and their "awakenings." No one had
ever commissioned me to write an article before, and *The Listener* had a very high reputation, so I felt honored and excited: it
would be my first opportunity to convey to a general audience
the wonder of the whole experience. And instead of the cen-
sorious rejections I had been getting from neurology journals,
I was actually being invited to write, being offered a chance to
publish fully and freely what had been accumulating and building up, dammed up, for so long.

The following morning I wrote the article at a single sitting
and messengered it to Mary-Kay. But by the afternoon, I had
second thoughts, and I phoned her to say that I felt I could do
a better job. She said the article I had sent her was fine, but if
I would like to make any additions or revisions, she would be
glad to read them. "But it does not *need* revision," she emphasized. "It is very clear, it flows easily—we would be happy to
publish it as it is."

But I felt I had not said everything I wanted to, and rather
than tamper with the original piece, I wrote another one, very
different in approach from the first. Mary-Kay was equally
pleased with this; both were publishable as they stood, she said.

By the next morning, I was again dissatisfied and wrote a
third draft, and that afternoon a fourth. Over the course of a
week, I sent Mary-Kay nine drafts in all. She then took off to
Scotland, saying she would try to conflate them somehow. She

returned after a few days, saying she found it impossible to combine them; each one had a different character, was written from a different perspective. They were not parallel versions, she said, they were "orthogonal" to one another. I would have to choose one, and if I could not, she would do so. She finally selected the seventh (or was it the sixth?) version, and that is what appeared in *The Listener* of October 26, 1972.

•

It seems to me that I discover my thoughts through the act of writing, *in* the act of writing. Occasionally, a piece comes out perfectly, but more often my writings need extensive pruning and editing, because I may express the same thought in many different ways. I can get waylaid by tangential thoughts and associations in mid-sentence, and this leads to parentheses, subordinate clauses, sentences of paragraphic length. I never use one adjective if six seem to me better and, in their cumulative effect, more incisive. I am haunted by the density of reality and try to capture this with (in Clifford Geertz's phrase) "thick description." All this creates problems of organization. I get intoxicated, sometimes, by the rush of thoughts and am too impatient to put them in the right order. But one needs a cool head, intervals of sobriety, as much as one needs that creative exuberance.

Like Mary-Kay, Colin had to pick among many versions, restrain my sometimes overabundant prose, and create a continuity. Sometimes he would say, pointing to one passage, "This doesn't go here," then flip the pages over, saying, "It goes *here*." As soon as he said this, I would see that he was right, but—mysteriously—I could not see it for myself.

It was not just unmuddling that I demanded of Colin at this time; it was emotional support when I was blocked or when my mood and confidence sagged, as they did, almost to the point of collapse, after the first rush was over.

September 19, 1972

Dear Mr. Haycraft,

I seem to be in one of those dry, dead depressed phases where one can only do nothing or blunder round in circles. The damn thing is that it needs only three days good work to finish the book, but I don't know whether I am capable of this at the moment.

I am in such an uneasy, guilt-stricken mood at the moment that I think I can't bear the thought of any of my patients being recognizably exposed, or the hospital itself being recognized in *Awakenings*—maybe this is one of the things which is inhibiting me from finishing the book.

It was now past Labor Day, America was back at work, and I too had to return to the daily grind in New York. I had finished another eleven cases, but I had no idea how to complete the book.

I returned to the familiar apartment next to Beth Abraham where I had been living since 1969, but the following month the director of the hospital told me abruptly that I had to get out: he needed the apartment for his ailing old mother. I said I appreciated her need, but it was my understanding that the apartment was reserved for the hospital's doctor on call and, as such, had been mine to occupy for the past three and a half years. My answer irked the director, who said that because I was questioning his authority, I could leave the apartment *and* the hospital. So, in a stroke, I was deprived of job, of income, of my patients, and of a place to live. (I continued, however, to visit my patients, albeit unofficially, until 1975, when I was formally reinstated at Beth Abraham.)

The apartment, which I had filled with my things, including a piano, looked desolate as it was stripped of everything, and I was in my now-empty apartment on November 13 when my brother David phoned me to say that our mother had died: she had had a heart attack during a trip to Israel and died while walking in the Negev.

I took the next plane to England and with my brothers carried her coffin at the funeral. I wondered how I would feel about sitting shiva. I did not know if I could bear it, sitting all day on a low stool with my fellow mourners for seven days on end, receiving a constant stream of people, and talking, talking, talking endlessly of the departed. But I found it a deep and crucial and affirmative experience, this total sharing of emotions and memories, when, alone, I felt so annihilated by my mother's death.

Only six months before, I had consulted Dr. Margaret Seiden, a neurologist at Columbia, after I had rushed up the cellar stairs in my apartment, hitting my head on a low beam and injuring my neck. After examining me, she asked whether my mother was a "Miss Landau." I said yes, and Dr. Seiden told me that she had been one of my mother's students; she was very poor at the time, and my mother paid her fees for medical school. It was only at Ma's funeral, when I met a number of her former students, that I learned how she had helped many of them through medical school, sometimes paying the entire cost. My mother had never told me (or anyone, perhaps) of the lengths to which she would go for her needy students. I had always thought of her as frugal, even parsimonious, but never realized how generous she was. I realized, too late, that there were whole sides to her which I had known nothing about.

My mother's older brother, my Uncle Dave (we called him Uncle Tungsten, and it was he who introduced me to chemistry as a boy), told me many stories of Ma's younger days, stories which fascinated me, comforted me, and sometimes made me laugh. Towards the end of the week, he said, "Come and have a good talk with me when you're back in England. I am the only one now who remembers your mother as a child."[8]

It especially moved me to see so many of my mother's patients and students and how they remembered her so vividly and humorously and affectionately—to see her through their eyes, as physician and teacher and storyteller. As they spoke of her, I was reminded of my own identity as a physician, teacher, and storyteller and how this had brought us closer, adding a new dimension to our relationship, over the years. It made me feel too that I must complete *Awakenings* as a last tribute to her. A strange sense of peace and sobriety, and of what really mattered, a sense of the allegorical dimensions of life and death, grew stronger and stronger in me with each day of the mourning.

My mother's death was the most devastating loss of my life—the loss of the deepest and perhaps, in some sense, the realest relation of my life. I found it impossible to read anything mundane; I could only read the Bible or Donne's *Devotions* when I finally went to bed each night.

---

8. When I returned to London a few months later, however, Uncle Dave himself was mortally ill. I visited him in his hospital room, but he was too weak now to talk at any length, so this, sadly, was a farewell visit to an uncle who had been so important, such a mentor, to me in my own boyhood, and I never did learn what my mother was like in *her* early days.

When the formal mourning was over, I stayed in London and returned to writing, with a sense of my mother's life and death and Donne's *Devotions* dominating all my thoughts. And in this mood, I wrote the later, more allegorical sections of *Awakenings*, with a feeling, a voice, I had never known before.

•

Colin unmuddled and soothed my moods, along with all the intricate, convoluted, sometimes labyrinthine ins and outs of the book so that by December it was finally finished. I could not bear the empty, motherless house at Mapesbury, and in the final month of writing I more or less moved to the Duckworth offices in the Old Piano Factory, though I would return to Mapesbury in the evenings to have dinner with Pop and Lennie (Michael, feeling psychosis rising again within him after Ma's death, had himself admitted to a hospital). Colin gave me a little room at Duckworth, and because my impulse to cross out or fiddle with what I had just written was so great at this time, we agreed that I would slip each page under the door as it was written. It was not just critical acumen that he provided but a sense of shelter and support, finally almost a home, which I needed quite as much at the time.

By December, then, the book was written.[9] The last page had been given to Colin, and it was time to go back to New

---

9. With the death of my mother and the completion of *Awakenings* (it was still untitled), I felt a peculiar compulsion to read and see Ibsen plays; Ibsen called to me, called to my condition, and his was the only voice I could bear.

Once I returned to New York, I went to every Ibsen play I could, but I could not find a performance of the play I most wanted to see, *When We Dead Awaken*. Finally, in the middle of January, I found that it was being performed at a small theater in northern Massachusetts and drove straight up to see it; the weather was nasty, and the smaller roads were treacherous. It was not the best of performances, but I identified with Rubek, the guilt-stricken artist. In that moment, I decided I had to title my own book *Awakenings*.

York. I took a taxi to the airport feeling that the book was complete. But then, in the taxi, I suddenly realized that something absolutely crucial had been omitted—something without which the entire structure would collapse. I hastily wrote it, and this was the beginning of a period of feverish footnote writing which continued for two months. This was long before the era of faxing, but by February of 1973 I had sent Colin more than four hundred footnotes by express mail.

Lennie had been in contact with Colin, who told her I was "fiddling" with the manuscript and deluging him with footnotes from New York, and this elicited a stern admonition from her: "*Don't, don't, don't* tamper with it or add any more footnotes!" she wrote.

Colin said, "The footnotes are all fascinating, but in aggregate they come to three times the length of the book, and they will sink it." I could only keep a dozen, he said.

"Okay," I replied. "You choose them."

But he said (wisely), "No, you choose them, because otherwise you'll be angry with me for my choice."

And so the first edition had only a dozen footnotes. Between them, Lennie and Colin saved *Awakenings* from my too-muchness.

I was thrilled to see galley proofs of *Awakenings* early in 1973. There were page proofs a couple of months later, but Colin never sent me these, because he was afraid that I would seize the opportunity to make innumerable changes and additions, as I had done with the galleys, and this would delay the scheduled publication.

Ironically, it was Colin who, a few months later, suggested postponing publication so that sections could be prepublished in *The Sunday Times*, but I was strongly against this, because

I wanted to see the book published on or before my birthday in July. I would be forty then, and I wanted to be able to say, "I may be forty, I have lost my youth, but at least I have done something, I have written this book." Colin thought I was being irrational, but seeing my state of mind, he agreed to stick to the original publication date in late June. (He later recollected that Gibbon had been at pains to publish the final volume of the *Decline and Fall* on *his* birthday.)

S taying at Oxford after my degree and often revisiting it in the late 1950s, I occasionally glimpsed W. H. Auden around town. He had been appointed a visiting professor of poetry at Oxford, and when he was there, he would go to the Cadena Cafe every morning to chat with anyone who wanted to drop by. He was very genial, but I felt too shy to approach him. In 1967, however, we met at a cocktail party in New York.

He invited me to visit, and I would sometimes go to his apartment on St. Mark's Place for tea. This was a very good time to see him, because by four o'clock he had finished the day's work but had not yet started the evening's drinking. He was a very heavy drinker, although he was at pains to say that he was not an alcoholic but a drunk. I once asked him what the difference was, and he said, "An alcoholic has a personality change after a drink or two, but a drunk can drink as much as he wants. I'm a drunk." He certainly drank a great deal; at dinner, either at his own place or someone else's, he would leave the meal at 9:30 p.m., taking all the bottles on the table with him. But however much he drank, he was up and at work by six the next morning. (Orlan Fox, the friend who introduced us, called him the least lazy man he had ever met.)

Wystan, like me, had grown up in a medical household. His father, George Auden, was a physician in Birmingham who served as medical officer during the great epidemic of encephalitis lethargica. (Dr. Auden was especially interested in how the disease could alter the personality in children, and he published several papers on this.) Wystan loved medical talk, and he had a soft spot for physicians. (In his book *Epistle to a Godson*, there are four poems dedicated to doctors, including one to me.) Knowing this, in 1969 I invited Wystan to visit Beth Abraham and meet my postencephalitic patients. (He later wrote a poem called "Old People's Home," but I have never been sure whether this was about Beth Abraham or some other home.)

He had written a lovely review of *Migraine* in 1971, and I was very excited by this; he was also critically important to me during the writing of *Awakenings*, especially when he said to me, "You're going to have to go beyond the clinical. . . . Be metaphorical, be mystical, be whatever you need."

·

By the beginning of 1972, Wystan had decided to leave America to spend his remaining days in England and Austria. He found the start of that winter a particularly grim one, with a mixed sense of illness and isolation, as well as the complex and contradictory feelings aroused by his decision to leave America, where he had lived so long and loved so deeply.

His first real break from this feeling came on his birthday, February 21. Wystan always loved birthdays and celebrations of all sorts, and this one was particularly important and moving. He was sixty-five; it was to be his last birthday in America, and his publishers had prepared a special party for him, where he was surrounded by an astonishing range of friends old and new (Hannah Arendt, I remember, sat next to him). It was only then, at this extraordinary gathering, that I fully realized the richness of Wystan's personality, his genius for friendship of all types. He sat beaming, ensconced in the middle of his friends, completely at home. Or so it seemed to me: I had never seen him happier. And yet, interfused with this, there was also a sense of sunset, of farewell.

Just before Wystan finally left America, Orlan Fox and I were helping him sort and pack his books, a painful task. After hours of shoving and sweating, we paused for a beer and sat without saying anything for a time. After a while, Wystan got up and said to me, "Take a book, some books, anything you want." He paused, then seeing my paralysis, he said, "Well, I'll decide then. These are *my* favorite books—two of them, anyhow!"

He handed me his libretto for *The Magic Flute* and a much-tattered volume of Goethe's letters, which he fetched from its place on his bedside table. The old Goethe was full of affectionate scribblings, annotations, and comments.[10]

At the end of the week—it was Saturday, April 15, 1972—Orlan and I ran Wystan to the airport. We arrived about three hours early, because Wystan was obsessively punctual and had an absolute horror of missing trains or planes. (He once told me of a recurrent dream of his: he was speeding to catch a train, in a state of extreme agitation; he felt his life, everything, depended on catching it. Obstacles arose, one after the other, reducing him to a silently screaming panic. And then, suddenly, he realized that it was too late, that he had missed the train, and that

---

10. He left his stereo and all of his records—a vast number of 78s as well as LPs—in New York, asking me if I would "look after them." I kept and played them for many years, though it got harder and harder to replace the tubes in the amplifier. In 2000, I gave them to the Auden archive at the New York Public Library.

it didn't matter in the least. At this point, there would come over him a sense of release amounting to bliss, and he would ejaculate and wake up with a smile on his face.)

We arrived early, then, and whiled away the hours in a meandering conversation; it was only later, when he left, that I realized that all the amblings and meanderings returned to one point: that the focus of the conversation was farewell—to us, to those thirty-three years, half of his life, which he had spent in the United States (he used to call himself a transatlantic Goethe, only half-jokingly). Just before the call for the plane, a complete stranger came up and stuttered, "You must be Mr. Auden. . . . We have been honored to have you in our country, sir. You'll always be welcome back here as an honored guest— and a friend." He stuck out his hand, saying, "Good-bye, Mr. Auden, God bless you for everything!" and Wystan shook it with great cordiality. He was much moved; there were tears in his eyes. I turned to Wystan and asked whether such encounters were common.

"Common," he said, "but never common. There is a genuine love in these casual encounters." As the decorous stranger discreetly retired, I asked Wystan how he experienced the world, whether he thought of it as being a very small or very large place.

"Neither," he replied. "Neither large nor small. Cozy, cozy." He added in an undertone, "Like home."

He said nothing more; there was no more to be said. The loud impersonal call blared out, and he hurried to the boarding gate. At the gate, he turned and kissed us both—the kiss of a godfather embracing his godsons, a kiss of benediction and farewell. He suddenly looked terribly old and frail but as nobly formal as a Gothic cathedral.

•

In February 1973, I was in England, and I went to Oxford to see Wystan, who by then had lodgings in Christ Church. I wanted to give him the galleys of *Awakenings* (he had asked for these, and in fact he was the only person who saw the galleys, other than Colin and Auntie Len). It was a beautiful day, and instead of taking a cab from the station, I decided to walk. I arrived somewhat late, and when I saw Wystan, he was swinging a watch. He said, "You're seventeen minutes late."

We spent a good deal of time discussing an article in *Scientific American* which had greatly excited him—Gunther Stent's "Prematurity and Uniqueness in Scientific Discovery." Auden had written a reply to Stent, contrasting the intellectual histories of science and art (this was published in the February 1973 issue).

Back in New York again, I got a letter from him. It was dated February 21—"my birthday," he added—short, and very, very sweet:

> Dear Oliver,
>
> Thanks so much for your charming letter. Have read *Awakenings* and think it a masterpiece. I do congratulate. My only query, if you want laymen to read it—as they should—is that you should add a glossary about the technical terms you use.
>
> Love,
>
> Wystan

I wept when I received Auden's letter. Here was a great writer, not given to facile or flattering words, judging my book "a masterpiece." Was this, however, a purely "literary" judgment? Was *Awakenings* of any *scientific* worth? I hoped so.

Later that spring, Wystan wrote to me again, saying his heart had been "acting up a bit" and that he hoped I could come to the house he shared with Chester Kallman in Austria. But I did not go, for one reason or another, and I deeply regret that I didn't visit him that summer, for he died on September 29.

On June 28, 1973 (*Awakenings*'s publication day), *The Listener* published a wonderful review of *Awakenings* by Richard Gregory and, in the same issue, my own article about Luria (I had been invited to review *The Man with a Shattered World* and to expand my review to include Luria's entire oeuvre). The following month, I was thrilled to receive a letter from Luria himself.

He later described how as a young man, a nineteen-year-old who had founded the grandiloquently titled Kazan Psychoanalytic Association, he had received a letter from Freud (who did not realize that he was writing to a teenager). Luria was hugely excited at receiving a letter from Freud, and I felt a similar excitement receiving one from Luria.

He thanked me for writing the article and dealt at length with all the points I had raised in it, indicating in very courteous but no uncertain terms that he thought I was deeply mistaken in various ways.[11]

A few days later, I got another letter in which Luria spoke of receiving the copy of *Awakenings* which Richard had sent him:

---

11. His letter then moved into a different mode and related an astonishing story of his meeting with Pavlov: the old man (Pavlov was then in his eighties), looking like Moses, tore Luria's first book in half, flung the fragments at his feet, and shouted, "You call yourself a scientist!" This startling episode was related by Luria with vividness and gusto in a way that brought out its comic and terrible aspects equally.

My dear Dr. Sacks,

I received *Awakenings* and have read it at once with great delight. I was ever conscious and sure that a good clinical description of cases plays a leading role in medicine, especially in Neurology and Psychiatry. Unfortunately, the ability to describe which was so common to the great Neurologists and Psychiatrists of the nineteenth century is lost now, perhaps because of the basic mistake that mechanical and electrical devices can replace the study of personality. Your excellent book shows, that the important tradition of clinical case studies can be revived and with a great success. Thank you so much for the delightful book!

A. R. Luria

I revered Luria as the founder of neuropsychology and of "romantic science," and his letter gave me great joy and a sort of intellectual reassurance I had never had before.

July 9, 1973, was my fortieth birthday. I was in London, *Awakenings* had just been published, and I was having a birthday swim in one of the ponds on Hampstead Heath, the pond in which my father had dunked me when I was a few months old.

I swam out to one of the buoys in the pond and was clinging to it, taking in the scene—there are few more beautiful places to swim—when I was groped underwater. I started violently, and the groper surfaced, a handsome young man with an impish smile on his face.

I smiled back, and we got talking. He was a student at Harvard, he told me, and this was his first time in England. He especially loved London, had been "seeing the sights" of the city every day and going to plays and concerts every evening. His nights, he added, had been rather lonely. He was due to return to the States in a week. A friend, now out of town, had lent him his flat. Would I care to visit?

I did so, happily, without my usual cargo of inhibitions and fears—happy that he was so nice looking, that he had taken the initiative, that he was so direct and straightforward, happy, too, that it was my birthday and that I could regard him, our meeting, as the perfect birthday present.

We went to his flat, made love, lunched, went to the Tate in the afternoon, to the Wigmore Hall in the evening, and then back to bed.

We had a joyous week together—the days full, the nights intimate, a happy, festive, loving week—before he had to return to the States. There were no deep or agonized feelings; we liked each other, we enjoyed ourselves, and we parted without pain or promises when our week was up.

It was just as well that I had no foreknowledge of the future, for after that sweet birthday fling I was to have no sex for the next thirty-five years.[12]

_____

12. In 2007, as I was starting a five-year stint as a professor of neurology at Columbia, I had to complete a medical interview to be cleared for working in the hospital. Kate, my friend and assistant, was with me, and at one point my interviewer, a nurse, said, "I have something rather private to ask you. Would you like Ms. Edgar to leave the room?"

"Not necessary," I said. "She is privy to all my affairs." I thought she was going to ask me about my sexual life, so without waiting for her question, I blurted out, "I haven't had any sex for thirty-five years."

"Oh, you poor thing!" she said. "We'll have to do something about that!" We all laughed; she had merely been going to ask me for my Social Security number.

Early in 1970, *The Lancet* had published my four letters to the editor about my postencephalitic patients and their responses to L-dopa. I assumed these letters would be read only by fellow physicians and was startled, a month later, when the sister of Rose R., one of my patients, held up a copy of the New York *Daily News,* which had reprinted, indeed highlighted, one of my letters under a headline.

"Is this your medical discretion?" she asked, waving the paper in front of me. Though only a close friend or a relative could have recognized the patient from the description, I was as shocked as she was—it did not occur to me that *The Lancet* would release an article to a news agency—I had thought professional writing had only a very limited circulation, not in the public sphere at all.

I had written a number of somewhat more technical papers in the mid-1960s—for journals like *Neurology* and *Acta Neuropathologica*—and there were no leaks to the news agencies then. But now, with my patients' "awakenings," I had entered a much broader arena, and this was my introduction to a very delicate, sometimes ambiguous area—a borderline or borderland between what can be said and what cannot be said.

I could not, of course, have written *Awakenings* without the encouragement and permission of the patients themselves, who had an overwhelming feeling of having been disposed of by society, put away, forgotten, and wanted their stories to be told. Nonetheless, I was hesitant, after the episode with the *Daily News,* to publish *Awakenings* in the United States. But one of my patients somehow got wind of the English publication and wrote to Colin, who sent her a copy of *Awakenings.* And then it was out.

•

Unlike *Migraine,* which had earned good opinions from both general reviewers and medical reviewers, the publication of *Awakenings* was greeted in a puzzling way. It was very well reviewed in the press generally. Indeed, it was awarded the 1974 Hawthornden Prize, a venerable award for "imaginative litera- ture." (I was thrilled by this, as I joined a list including Robert Graves and Graham Greene, among others—to say nothing of James Hilton for *Lost Horizon,* a book I adored as a boy.)

But there was not a murmur from my medical colleagues. No medical journals reviewed it. Finally, in January of 1974, the editor of a rather briefly lived journal called *The British Clinical Journal* wrote that he thought that two of the strang- est phenomena in England in the preceding year had been the publication of *Awakenings* and the complete lack of medical response to it, what he called the "strange mutism" of the profession.[13]

Nonetheless, the book was voted Book of the Year by five eminent writers, and in December of 1973 Colin threw a com- bined publication and Christmas party. There were many peo- ple at the party I had heard of and admired but had never met or thought to meet. My father, who was just recovering from a year of mourning for my mother, came to the party, and he, who had been so anxious about my publishing, saw all sorts of eminent people there and was greatly reassured. I myself, who had felt so lost, so unknown, now felt rather lionized and

---

13. It was not until some years later that the strange, unstable states I saw with my postencephalitic patients were observed in "ordinary" par- kinsonian patients maintained on L-dopa. These patients, with their more stable nervous systems, might not show such effects for several years (while the postencephalitics developed them within weeks or months).

feted. Jonathan Miller was at the party too, and he said to me, "You're famous now."

I didn't really know what that meant; no one had ever said anything like this to me before.

•

I had one review in England which irked me, although it was quite positive in most ways. I had of course given the patients pseudonyms in the book and had given Beth Abraham a pseudonym as well. I called it Mount Carmel and located it in the fictitious village of Bexley-on-Hudson. This reviewer wrote something like, "This is an amazing book, the more so since Sacks is talking about non-existent patients in a non-existent hospital, patients with a non-existent disease, because there was no worldwide epidemic of sleepy sickness in the 1920s." I shared this review with some of the patients, and many of them said, "*Show* us, or the book will never be believed."

And so I asked all the patients how they would feel about a documentary. They had encouraged me, earlier, about publishing the book: "Go ahead; tell our story—or it will never be known." And now they said, "Go ahead; film us. Let us speak for ourselves."

I was not sure of the propriety of showing my patients on film. What passes between physician and patient is confidential and even to write of it, in some sense, is a breach of this confidence, but writing allows one to change names and places and certain other details. Such disguise is impossible in a documentary film; faces, voices, real lives, identities, are all exposed.

So I had misgivings, but I was approached by several documentary film producers and was particularly impressed by one of them, Duncan Dallas of Yorkshire Television, especially by his combination of scientific knowledge and human feeling.

Duncan came to Beth Abraham for a visit in September of 1973 and met all of the patients. Many he recognized from having read their stories in *Awakenings*. "I know you," he said to several of them. "I feel I've met you before."

He also asked, "Where's the music therapist? She seems to be the most important person around here." He was referring to Kitty Stiles, an unusually talented music therapist. It was quite unusual, in those days, to *have* a music therapist—the effects of music, if any, were considered no more than marginal—but Kitty, working at Beth Abraham since the early 1950s, knew that patients of all sorts could respond strongly to music and that even the postencephalitics, although often incapable of initiating movements voluntarily, could respond to a beat involuntarily, as we all do.[14]

Almost all of the patients warmed to Duncan and realized he would present them with objectivity and a discreet compassion, neither over-medicalizing nor over-sentimentalizing their lives. When I saw how quickly a mutual understanding and respect was established, I agreed to the filming, and Duncan returned with his crew the following month. Some of the patients, of course, did not want to be filmed, but most of them felt it was important to show themselves as human beings who had been forced to dwell in a deeply strange world.

---

14. By 1978, Kitty had decided to retire; we thought she had reached the usual retirement age of sixty-five, but she was, we learned, in her nineties, though astonishingly youthful and vivacious (could music have kept her young?). Kitty's replacement was Connie Tomaino, an energetic young woman with an advanced degree in music therapy who would go on to organize a massive, wide-ranging program of music therapy, exploring what musical approaches were most suitable for patients with dementia, patients with amnesia, and patients with aphasia. Connie and I collaborated for many years, and she is still at Beth Abraham, now as director of the Institute for Music and Neurologic Function.

Duncan incorporated some of the Super 8 film I had shot in 1969 showing the patients' awakenings as they were given L-dopa and then as they suffered bizarre tribulations of all types, and he added moving interviews with the patients as they looked back on these events and described how they were now living their lives after having been out of the world for so many years.

The documentary film of *Awakenings* was broadcast in England early in 1974. It is the only documentary account of these last survivors of a forgotten epidemic and how their lives were transformed for a while by a new drug; of how intensely human they were, throughout all their vicissitudes.

The Bull on the Mountain

After my mother's death, I returned to a wintry New York. Having just been fired by Beth Abraham, I had no apartment, no real job, and no significant income.

I had, however, been working as a consultant doing a weekly neurology clinic at the Bronx Psychiatric Center, familiarly known as Bronx State. I would examine patients, usually diagnosed as schizophrenic or manic-depressive, to see whether they might have some neurological condition as well. Like my brother Michael, patients on tranquilizers often developed movement disorders (parkinsonism, dystonia, tardive dyskinesia, etc.), and these movement disorders often persisted long after the medications were stopped. I spoke to many patients who said that they could live with their mental disorders but not with the movement disorders we had given them.

I also saw patients whose psychoses or schizophrenia-like conditions were due to (or augmented by) neurological diseases. I recognized several undiagnosed or misdiagnosed post-encephalitic patients in Bronx State's back wards and found others with brain tumors or degenerative brain diseases.

But this job occupied only a few hours a week and paid very little. Seeing my situation, the director at Bronx State, Leon Salzman (a very genial man who had written an excellent book on the obsessional personality), invited me to work half-time in the hospital. He thought I would be particularly interested in Ward 23, a ward where young adults with a variety of problems—autism, retardation, fetal alcohol syndrome, tuberous sclerosis, early-onset schizophrenia, etc.—were warehoused together.

Autism was not a hot subject at that time, but it was one which interested me, so I accepted the offer. At first, I enjoyed being on this ward, although it upset me deeply as well. Neurologists, perhaps more than any other specialists, see tragic cases—people with incurable, relentless diseases which can cause great suffering. There has to be, along with fellow feeling and sympathy and compassion, a sort of detachment so that one is not drawn into a too-close identification with patients.

But Ward 23 had a so-called behavior modification policy, using rewards and punishments, and in particular "therapeutic punishment." I hated to see the way in which the patients were treated, sometimes locked in seclusion rooms or starved or restrained. Among other things, it reminded me of the way I had been treated as a child when I was sent away to a boarding school where I (and other boys) were frequently punished by a capricious and sadistic headmaster. I felt myself falling sometimes into an almost helpless identification with the patients.

I observed these patients closely, felt for them, and tried, as a physician, to bring out their positive potentials. I tried to engage them, whenever possible, in the morally neutral realm of play. With John and Michael, autistic and retarded twins who were calendar and number savants, play took the form of looking for factors or prime numbers; for José, a graphically gifted autistic boy, play was in the realm of drawing and visual arts; while for Nigel—a speechless, autistic, and probably retarded youth—music was crucial. I had my old upright piano transported to Ward 23, and when I played, Nigel and some of the other young patients would gather round the piano. Nigel, if he liked the music, would do strange and elaborate dances. (I spoke of him in one consultation note as "an idiot Nijinsky.")

Steve, also mute and autistic, was drawn to a pool table which I had found in the hospital basement and had moved up to the ward. He acquired skill with amazing speed, and although he spent hours alone at the table, he clearly enjoyed playing pool with me. This was, as far as I could see, his only social or personal activity. When not absorbed by the pool table, he was hyperactive, darting around, always in motion, lifting things up and examining them—a sort of exploratory behavior, half-compulsive, half-playful, as one may sometimes see in Tourette's syndrome or some frontal lobe disorders.

I was fascinated by these patients and started to write about them early in 1974. By April, I had completed twenty-four pieces—enough, I thought, for a small book.

Ward 23 was a locked ward, and being locked in was particularly hard on Steve. He would sometimes sit by the window or by the wire-glass-paneled door, longing to be outside. The staff never took him out. "He'll run off," they would say. "He'll make his escape."

I felt very sorry for Steve, and though he could not speak, I felt, from the way he would look for me and attach himself to me at the pool table, that he would not run away from me. I spoke to a colleague—a psychologist at Bronx Developmental Services, a day program where I also did a weekly session—and he agreed, after meeting Steve, that the two of us could safely take him out together. We broached the idea to Dr. Taketomo, the unit chief on Ward 23, who thought about it carefully and then agreed, saying, "He's your responsibility if you take him out. Make sure he comes back safe and sound."

Steve was startled when we took him out of the ward but seemed to understand that we were going on an outing. He got into the car, and we drove to the New York Botanical Garden, ten minutes from the hospital. Steve loved the plants; it was May and the lilac was in full bloom, He loved the grassy dells and the spaciousness all about him. At one point, he picked a flower, gazed at it, and uttered the first word any of us had ever heard him say: "Dandelion!"

We were stunned; we had no idea that Steve could recognize any flowers, let alone name them. We spent half an hour in the garden and then drove back slowly so Steve could get a good look at the crowds and shops on Allerton Avenue, the bustle of life he was so cut off from in Ward 23. He resisted a little as we walked back into the ward but seemed to understand there might be further outings.

The staff, who had been unanimously opposed to the excursion and had predicted it would end in disaster, seemed furious at our descriptions of Steve's good behavior and obvious happiness in the garden and his uttering his first word. We were greeted with black looks.

I had always made a point of avoiding the big Wednesday staff meetings, but the day after our outing with Steve, Dr. Taketomo insisted that I come. I felt apprehensive of what I might hear, and even more of what I might say. And my apprehensions were fully justified.

The chief staff psychologist said that a well-organized and successful behavior modification program had been set up and that I was undermining this by my notions of "play" not conditional on external rewards or punishment. I replied, defending the importance of play and criticizing the reward-punishment model. I said I thought this constituted a monstrous abuse of the patients in the name of science and sometimes smacked of sadism. My reply was not received too kindly, and the meeting ended in angry silence.

Two days later, Taketomo came up to me and said, "Rumors are going around that you are sexually abusing your young patients."

I was shocked and replied that such a thought would never enter my mind. I regarded patients as my charges, my responsibility, and I would never make use of my power as a therapeutic figure to exploit them.

In a growing rage, I added, "You may know that Ernest Jones, Freud's colleague and biographer, worked with retarded and disturbed children in London as a young neurologist until rumors came out that *he* was abusing his young patients. These rumors forced him out of England, and he then went to Canada."

He said, "Yes, I know. I wrote a biography of Ernest Jones."

I wanted to turn on him and say, "You fucking idiot, why did you get me into this?" But I didn't; he probably felt that he was just mediating a civilized discussion.

I went to Leon Salzman and told him the situation; he was sympathetic and angered on my behalf, but he thought it would be best for me to leave Ward 23. I felt an overwhelming if irrational guilt at abandoning my young patients, and on the night of my departure I threw the twenty-four pieces I had written into the fire. I had read that Jonathan Swift, in a desperate mood, had thrown the manuscript of *Gulliver's Travels* into the fire and that his friend Alexander Pope had pulled it out. But I was alone and had no Pope to pull my book out.

The day after I left, Steve escaped from the hospital and climbed up high on the Throgs Neck Bridge; mercifully, he was rescued before he could jump. This made me realize that my sudden, forced abandonment of my patients was at least as hard for them, and as dangerous, as it was for me.

I left Ward 23 foaming with guilt, remorse, and rage: guilt at leaving the patients, remorse for destroying the book, and rage at the accusations of abuse. They were false, but they made me deeply uncomfortable, and I thought that what I had delivered so fatally in a few words about the running of the ward in that Wednesday meeting, I would now expose to the world in a denunciatory book I would write, called "Ward 23."

I took off for Norway soon after my departure from Ward 23, because I thought that it would be a good, peaceful place to write my diatribe. But I had a series of accidents one after the other, gradually getting more and more serious. First I rowed far out on Hardangerfjord, one of the larger fjords in Norway, and then clumsily lost an oar overboard. Somehow I made my way back with one oar, but it took several hours, and I wondered once or twice whether I would make it.

The next day I started out for a little mountain walking. I was alone and had told no one where I was going. I saw a sign in Norwegian at the bottom of the mountain which said "Beware of the Bull"; it included a little cartoon of a man being tossed by a bull. I thought this must be the Norwegian sense of humor. How could you keep a bull on a mountain?

I dismissed it from my mind, but a few hours later, coming nonchalantly around a big boulder, I found myself face-to-face with a huge bull sitting squarely on the path. "Terror" is too mild a word for what I felt, and my fear induced a sort of hallucination: the bull's face seemed to expand until it filled the universe. Very daintily, as if I had casually decided to end my walk at this point, I turned around and started retracing my steps. But then my nerve broke, panic took over, and I started to run down the muddy, slippery path. I heard heavy, thudding footsteps and heavy breathing behind me (was the bull in pursuit?), and suddenly—I do not know how it happened—I was at the bottom of a cliff with my left leg twisted grotesquely underneath me.

One can have dissociations in times of extremity. My first thought was that someone, someone I knew, had had an accident, a bad accident, and only then did I realize that *I* was that someone. I tried to stand up, but the leg gave way like a strand of spaghetti, completely limp. I examined the leg—very professionally, imagining that I was an orthopedist demonstrating an injury to a class of students: "You see the quadriceps tendon has torn off completely, the patella can be flipped to and fro, the knee can be dislocated backwards: so." With that, I yelled. "This causes the patient to yell," I added, and then again came back to the realization that I was not a professor demonstrating an injured patient; I *was* the injured person. I had been using an umbrella as a walking stick, and now, snapping off the handle, I splinted the stem of the umbrella to my leg using strips of cloth I tore from my anorak and started my descent, levering myself down with my arms. At first I did so very quietly, because I thought the bull might still be in the vicinity.

I went through many different moods as I levered myself and my useless leg down the path. I did not see my life in a flash, but many, many memories unreeled. They were nearly all good memories, grateful memories, memories of summer afternoons, memories of having been loved, memories of having been given things, and gratitude that I had also given something back. In particular, I thought, I had written one good book and one great book; I found myself using the past tense. A line from an Auden poem, "Let your last thinks all be thanks," kept going through my mind.

Eight long hours passed, and I was in near-shock, with a considerable amount of swelling in the leg, though fortunately no bleeding. Soon it would be dark; the temperature was already going down. No one was searching for me; no one even knew where I was. Suddenly I heard a voice. I looked up and saw two figures on a ridge—a man holding a gun and a smaller figure next to him. They came down and rescued me, and I thought then that being rescued from almost certain death must be one of the sweetest experiences in life.

·

I was flown to England and, forty-eight hours later, operated on to repair the torn quadriceps tendon and muscle. But following the surgery, for two weeks or more, I could neither move nor feel the damaged leg. It felt alien, not a part of me, and I was deeply puzzled, confounded. My first thought was that I had suffered a stroke while under anesthesia. My second thought was that this was a hysterical paralysis. I found myself unable to communicate my experience to the surgeon who operated on me; all he could say was "Sacks, you're unique. I never heard anything like this before!"

Eventually, as the nerves recovered, the quadriceps came back to life: first in the form of fasciculations, individual bundles of muscle fibers twitching in the previously inert and toneless muscle; then as an ability to make small voluntary contractions of the quadriceps, to tense the muscle (where it had been jellylike, impossible to contract, for the previous twelve days); and finally in the ability to flex the hip, though the movement was erratic, weak, and easily fatigued.

At this stage, I was taken down to the casting room to have the cast changed and the stitches taken out. When the cast was removed, the leg looked quite alien, not "mine," more like a beautiful wax model from an anatomy museum, and I felt nothing whatever as the stitches were taken out.

After a new cast was put on, I was taken to the physiotherapy department to be stood up and walked. I use this odd, passive construction—"*to be* stood up and walked"—because I had forgotten *how* to stand and walk, do this actively, on my own. Hoisted to my feet and trying to stand, I was assailed by rapidly fluctuating images of my left leg: it seemed very long, very short, very slim, very squat. These images modulated to relative stability in a minute or two, my proprioceptive system recalibrating, I imagined, to the rush of sensory input and the first, sputtering motor output in a leg which had been without sensation or movement for two weeks. But moving the leg felt like manipulating a robot limb—consciously, experimentally, one step at a time. It was nothing like normal, fluent walking. And then, suddenly, I "heard," with hallucinatory force, a gorgeous, rhythmic passage from Mendelssohn's Violin Concerto. (Jonathan Miller had given me a tape of this when I went into hospital, and I had played it constantly.) With this playing in my mind, I found myself suddenly able to walk, to regain (as neurologists say) the "kinetic melody" of walking. When the inner music stopped after a few seconds, I stopped too; I needed the Mendelssohn to keep going. But within an hour, I had regained fluent, automatic walking and no longer needed my imaginary musical accompaniment.

Two days later, I was moved to Caenwood House—a baronial convalescent home on Hampstead Heath. My month there was an unusually social one. I was visited not only by Pop and

Lennie but by my brother David (who had arranged my flight from Norway and emergency admission to hospital in London) and even Michael. Nieces, nephews, and cousins came, neighbors, people from shul, and, almost daily, my old friends Jonathan and Eric. All this, combined with the sense that I had been saved from death and was recovering mobility and independence daily, gave a peculiarly festive quality to my weeks at the convalescent home.

Pop would sometimes visit after his morning consulting hours (though he was almost eighty, he still had a full working day). He made a point of visiting some of the elderly parkinsonian patients at Caenwood and would sing World War I–era songs with them; many of them, though they could hardly speak, were able to sing along once my father got them going. Lennie would come in the afternoons, and we would sit outside in the mild October sun and chat for hours. When I got more mobile and graduated from crutches to a stick, we would walk to local tea shops in Hampstead or Highgate Village.

The leg incident taught me in a way which I could not, perhaps, have learned otherwise about how one's body and the space around one are mapped in the brain and how this central mapping can be profoundly deranged by damage to a limb, especially if this is combined with immobilization and encasement. It also gave me a feeling of vulnerability and mortality which I had not really had before. In my earlier days on the motorbike, I was audacious in the extreme. Friends observed that I seemed to think of myself as immortal or invulnerable. But after my fall and my near death, fear and caution entered my life and have been with me, for better or worse, ever since. A carefree life became a careful one, to some extent. I felt this was the end of youth and that middle age was now upon me.

Almost as soon as the accident happened, Lennie perceived that there was a book to be written about it, and she liked to see me, pen in hand, writing in my notebook. ("Don't use a ballpoint!" she admonished me sharply; her own beautifully legible, rounded handwriting was always done with a fountain pen.)

Colin was alarmed when he heard of my accident but fascinated when I told him how it had happened and what was going on with me in the hospital. "This is grand stuff!" he exclaimed. "You have to write all about it." He paused and then added, "It sounds as if you're actually living the book right now." A few days later, he brought me an enormous dummy for a book he had just published (a dummy has no text, just the book's cover enclosing blank pages)—seven hundred empty, creamy white pages—so that I could write as I lay in my hospital bed. I was delighted with this huge notebook, the largest I had ever had, and kept very full notes of my own involuntary journey, as I saw it, into neurological limbo and back. (Other patients, seeing me with this huge book, would say, "You lucky bugger—we're just going through it, but you're making a book of it.") Colin called frequently to check on my progress—the progress of my "book," quite as much as my progress as a patient—and his wife, Anna, often came as well, bringing gifts of fruits and smoked trout.

The book I wanted to write would be about the losing and the reclaiming of a limb. Since I had called my last book *Awakenings,* I thought I would call this next one "Quickenings."

But there were to be problems with this book of a sort I had not had before, because writing it involved reliving the accident, reliving the passivity and horrors of patienthood; it involved, too, an exposure of some of my own intimate feelings in a way which my more "doctorly" writings had never done.

There were many other problems. I had been elated, and a bit daunted, by the response to *Awakenings*. Auden and others had said what I had hardly dared to think—that *Awakenings* was a major work. But if this was so, I saw no way of following it with anything comparable. And if *Awakenings*, with its wealth of clinical observation, had been ignored by my colleagues, what could I expect of a book entirely concerned with the odd and subjective experience of a single subject—myself?

By May of 1975, I had written a first rough draft of "Quickenings" (later to be titled, at Jonathan Miller's suggestion, "A Leg to Stand On"). I felt, as Colin did, that it could soon be readied for publication. Colin was so confident, indeed, that he included it in his upcoming 1976–77 catalogue.

But something went wrong between Colin and me that summer of 1975, as I strove to finish the book. The Millers went up to Scotland in August and allowed me the use of their house in London. This was right opposite Colin's house, as close as one could get—what could be more ideal for the work that lay ahead? But the proximity which had been so delightful, so productive with *Awakenings* now unhappily had the opposite effect. I would write every morning, spend the afternoon walking or swimming, and every evening, around seven or eight, Colin would come by. He had eaten by then, and usually drunk a good deal too, and tended more often than not to be flushed, irritable, and argumentative. The August nights were hot and airless, and perhaps there was something about my manuscript or something about me which brought out his anger; I was tense and anxious that summer and uncertain about my writing. He would pick one of the pages I had typed, read a sentence or a paragraph, and then attack it—its tone, its style, its substance. He would take each sentence, each thought, and worry it to death—or so it seemed to me. He showed, I thought, none of the humor, the geniality, I had expanded in before but a censoriousness so strict I shriveled before it. After these evening sessions, I would have an impulse to tear up the day's work, to feel the book was idiocy—that I could or should not go on.

The summer of 1975 ended on an evil note, and (though I never encountered Colin in such a state ever again) it cast a shadow on the years to come. Thus *Leg* was not, after all, completed that year.

Lennie worried about me: *Awakenings* was finished, *Leg* was running into difficulties, and I did not seem to have any special project to animate me. She wrote, "I do so hope . . . that the sort of work that's right for you will come your way, and continue to do so. I feel strongly that you *must* write, whether you are in the mood for it or not." Two years later, she added, "*Do* get the leg book off your mind and write your next."

•

Many versions of *Leg* were to be written over the next few years, each longer, more intricate, more labyrinthine than the last. Even the letters I sent to Colin were of inordinate length— one, from 1978, ran to more than five thousand words, with an addendum of another two thousand.

I also wrote to Luria, who replied patiently and thoughtfully to my overlong letters. Finally, when he saw me obsessing endlessly about a possible book, he sent me a two-word telegram: "DO IT."

He followed this with a letter in which he spoke of the "central resonances of a peripheral injury." He went on, "You are discovering an entirely new field. . . . Please publish your observations. It may do something to alter the 'veterinary' approach to peripheral disorders, and to open the way to a deeper and more human medicine."

But the writing—the incessant writing and tearing up of drafts—continued. I found *Leg* more painful and difficult than anything I had ever written, and some of my friends (Eric in particular), seeing me so obsessed and so stuck, urged me to give the book up as a bad job.

In 1977, Charlie Markham, a former mentor in neurology at UCLA, visited New York. I was fond of Charlie and had spent time with him when he was doing research on movement disorders. Over lunch, he quizzed me about my work and exclaimed, "But you have no position!"

I said I *did* have a position.

"What? What sort of position do you have?" he asked (he himself had recently ascended to chair of neurology at UCLA).

"At the heart of medicine," I answered. "That's where I am."

"Pfft," said Charlie, with a brief, dismissive gesture.

I had come to feel this during the years of my patients' "awakenings," when I lived next door to the hospital and sometimes spent twelve or fifteen hours a day with them. They were welcome to visit me; some of the more active ones would come over to my place for a cup of cocoa on Sunday mornings, and some I would take to the New York Botanical Garden, just opposite the hospital. I monitored their medications, their often unstable neurological states, but I did my best, too, to see that they had full lives—as full as possible, given their physical limitations. I felt that trying to open up the lives of these patients, who had been immobilized and shut up in hospital for so many years, was an essential part of my role as their physician.

Though I no longer had a position or salary at Beth Abraham, I continued to go there regularly. I was too close to my patients to allow any break in our contact, even though I started to see patients in other facilities—nursing homes all over New York City, from Staten Island to Brooklyn and Queens.[1] I became a peripatetic neurologist.

In some of these places, generically referred to as "the manors," I saw the complete subjugation of the human to medical arrogance and technology. In some cases, the negligence was willful and criminal—patients left unattended for hours or even abused physically or mentally. In one "manor," I found a patient with a broken hip, in intense pain, ignored by the staff and lying in a pool of urine. I worked in other nursing homes where there was no negligence but nothing beyond basic medical care. That those who entered such nursing homes needed meaning—a life, an identity, dignity, self-respect, a degree of autonomy—was ignored or bypassed; "care" was purely mechanical and medical.

I found these nursing homes as horrible in their way as Ward 23, and perhaps even more troubling, since I could not help wondering whether they represented portents or "models" of the future.

I found the exact opposite of the "manors" in the residential homes of the Little Sisters of the Poor.

---

1. In the late 1970s and early 1980s, I also spent some time in an Alzheimer's clinic at Einstein, and I prepared five long case histories based on some of these patients. I sent the manuscript of those to my former chief at Einstein, Bob Katzman (he had gone on to chair the neurology department at UCSD). But somehow, in the midst of moving, it got lost—another book, like "Myoclonus," which would never see the light of day.

I first heard of the Little Sisters when I was a boy, for both my parents consulted at their homes in London—my father as a general practitioner and my mother as a surgical consultant. Auntie Len would always say, "If I get a stroke, Oliver, or get disabled, get me to the Little Sisters; they have the best care in the world."

Their homes are about life—living the fullest, most meaningful life possible given their residents' limitations and needs. Some of the residents have had strokes, some have dementia or parkinsonism, some have "medical" conditions (cancer, emphysema, heart disease, etc.), some are blind, some are deaf, and others, though in robust health, have become lonely and isolated and long for the human warmth and contact of a community.

Besides medical care, the Little Sisters provide therapy of every kind—physical therapy, occupational therapy, speech therapy, music therapy, and (if need be) psychotherapy and counseling. In addition to therapy (though no less therapeutic) are activities of all sorts, activities which are not invented but real, like gardening and cooking. Many residents have special roles or identities in the homes—from helping in the laundry to playing the organ in the chapel—and some have pets to care for. There are outings to museums, racetracks, theaters, gardens. Residents with families may lunch out on weekends or stay with their relatives over the holidays, and the homes are visited regularly by children from nearby schools, who interact spontaneously and unself-consciously with people seventy or eighty years their senior and may form affectionate bonds with them. Religion is central but not mandatory; there is no preachiness, no evangelism, no religious pressure of any sort. Not all the residents are believers, though there is a great religious devotion among the Sisters, and it is difficult to imagine such a level of care without such a deep dedication.[2]

There may (and perhaps must) be a difficult period of adjustment in giving up one's own home for a communal one, but the vast majority of those admitted to the Little Sisters' homes are able to establish meaningful and enjoyable lives of their own—sometimes more so than they have known for years—along with the assurance that all their medical problems will be sensitively monitored and treated and that, when the time comes, they can die in peace and dignity.

All this represents an older tradition of care, preserved by the Little Sisters since the 1840s and, indeed, going back to the ecclesiastical traditions of the Middle Ages (such as Victoria Sweet describes so movingly in *God's Hotel*)—combined with the best that modern medicine can offer.

Though I was dispirited by the "manors" and soon stopped going to them, the Little Sisters inspire me, and I love going to their residences. I have been going to some of them, now, for more than forty years.

Early in 1976, I received a letter from Jonathan Cole, a medical student at the Middlesex in London. He spoke of having

---

2. Not infrequently dilemmas of an unusual sort arise, and here the Little Sisters show a moral breadth and clarity of mind. One of their residents, Flora D., a parkinsonian woman, was greatly helped by L-dopa but concerned by the extremely vivid dreams she started to get. It is not uncommon to have erotic dreams or nightmares on L-dopa, but Flora had incestuous dreams, of intercourse with her father. She felt guilty and extremely anxious about this until she described the dreams to one of the nuns, who said, "You are not responsible for the dreams you have at night. It would be quite different if these were *day*dreams." This was a clear moral distinction consonant with a clear physiological distinction.

enjoyed *Migraine* and *Awakenings* and added that he had done a year of research in sensory neurophysiology at Oxford before going on to clinical work. He wondered if he could spend his elective period, about two months, with me. "I would like," he wrote, "to observe the methods of your department and would willingly fit in with whatever teaching course exists."

I felt warmed, flattered, at being approached by a student at the hospital where I had been a medical student nearly twenty years earlier. But I had to disabuse him of various notions regarding my position and ability to provide the sort of teaching one has in medical school, so I wrote back to him:

Dear Mr. Cole,

I thank you for your letter of February 27, and am sorry to be so long in replying.

My delay in replying is because I don't know what to reply. But here, roughly, is my situation:

I don't *have* a Department.

I am not *in* a Department.

I am a gypsy, and survive—rather marginally and precariously—on odd jobs here and there.

When I worked full-time at Beth Abraham I often had students spend some time with me for their electives—and this was an experience we would always find very pleasant and rewarding.

But now I am, as it were, without any position or base or home, but peripatetic here and there. I can't possibly offer any *formal* sort of teaching—or anything which could be formally accredited to you.

Informally (I sometimes think) I see and learn and do a great deal, with the extremely varied patients I see in various clinics and Homes, and every seeing-and-learning-and-doing situation is, *eo ipso*, a teaching situation. I find every patient I see, everywhere, vividly alive, interesting and rewarding; I have never seen a patient who didn't teach me something new, or stir in me new feelings and new trains of thought; and I think that those who are with me in these situations share in, and contribute to, this sense of adventure. (I regard all neurology, *everything*, as a sort of adventure!)

Do write and let me know how things work out with you— once again, I would be delighted to see you in an informal, casual, peripatetic way, but I am in no sense "set up" for any formal teaching whatever.

With best wishes—and thanks,

Oliver Sacks

It took almost a year to make arrangements and provide funds, but early in 1977 Jonathan arrived for his elective with me.

We were both, I think, a little nervous: I was, after all, the author of *Awakenings*, even if I had no position, and Jonathan had done research in sensory neurophysiology at Oxford and was obviously far more sophisticated and up to date in physiological thinking than I was. It was going to be a new, unprecedented experience for us both.

We soon discovered one strong interest that we shared; we were both fascinated by "the sixth sense," proprioception: unconscious, invisible, but arguably more vital than any or all of the other five senses put together. One could be blind and deaf, like Helen Keller, and still lead a fairly rich life, but proprioception was crucial for the perception of one's own body, the position and movement of one's limbs in space, crucial

indeed for the perception of their *existence*. How, if proprioception were extinguished, would a human being survive?

Such a question would hardly arise in the usual course of
life; proprioception is always there, never intruding itself, but
quietly guiding every movement we make. I am not sure that
I would have thought about proprioception much had I not
had the bizarre disturbance which (at the very time Jonathan
came to New York) I was struggling to write about in my *Leg*
book—a disturbance stemming in large part, so I thought, from
the breakdown of proprioception, a breakdown so profound
that I could not say, without looking, where my left leg was, or
that it was, nor did it feel "mine."

And, by coincidence, around the time Jonathan came to
New York, my friend and colleague Isabelle Rapin sent me a
patient, a young woman, who, as a result of a viral illness, had
suddenly lost all proprioception and sense of touch from the
neck down.[3] Jonathan could not have known, in 1977, how
profoundly his life would be mingled in the future with that of
another patient who had the same condition.

Going with me to the Little Sisters and other homes all over
New York, Jonathan saw a variety of patients. One especially
stayed in both of our minds, a man with Korsakoff's syndrome
whose lack of memory forced him to confabulate continually.
In the course of three minutes, "Mr. Thompson" (as I later
called him) identified me (in my white doctor's coat) as a customer in his delicatessen, an old friend he would go to the races
with, a kosher butcher, and a gas station attendant; only then,

---

3. Some years later, under the title "The Disembodied Lady," I related
her story in *The Man Who Mistook His Wife for a Hat*.

with some prompting, did he guess that I might be his doctor.[4] I burst out laughing as he moved from one comic misrecognition or confabulation to another, and sober Jonathan (as he told me later) was shocked at this, shocked that I seemed to be laughing at a patient. But when Mr. Thompson, an ebullient Irishman, started laughing too, laughing at the antics of his own Korsakoffian imagination, Jonathan relaxed and started laughing himself.

I used to carry a video camera with me when I went to see patients, and Jonathan was intrigued by the uses of video recording and instant playback; video recording was rather new in those days and rarely used in hospitals. He was fascinated to see how patients with parkinsonism, for example, unaware of their tendency to accelerate or to lean to one side, could become aware of it through viewing their own postures or gait on video—and learn measures to correct these.

I took Jonathan to Beth Abraham several times; he was particularly keen to meet the patients he had read about in *Awakenings*. He was greatly intrigued, he told me, that it had been possible for me to write about these patients and even film them, yet continue to be seen by them as a trustworthy physician, not as someone who had exploited or betrayed them. This must have been very much in Jonathan's mind when, eight years later, he met Ian Waterman, the man who was to change his life.

Ian, like Christina—the disembodied lady—had suffered a devastating sensory neuropathy. He had been a robust nineteen-

---

4. I described Mr. Thompson in "A Matter of Identity" in *The Man Who Mistook His Wife for a Hat.*

year-old when a virus suddenly deprived him of all proprioception below his head. Most people in this rare situation are scarcely able to control their limbs at all and are confined to crawling or sitting in a wheelchair. But Ian had found many amazing ways of dealing with his condition and was able to lead a fairly normal existence, despite his profound neurological deficits.

Much that is automatic in the rest of us, occurring without the need for conscious supervision, Ian can only undertake with conscious deliberation and monitoring. When he sits, he must consciously hold himself erect so he does not fall forward; he can walk only if he locks his knees and keeps his eyes on the task. Lacking the "sixth sense" of proprioception, he must substitute vision instead. This focus and concentration means that he cannot easily do two things at once. He can stand, or he can talk, but in order to stand *and* talk, he must lean against a support. He may appear perfectly normal, but if the lights suddenly go out without warning, he will fall helplessly to the ground.

Over the years, Jonathan and Ian have formed a deep relationship—as doctor and patient, investigator and subject, and, increasingly as colleagues and friends (they have been working together for thirty years now). In the course of this decades-long collaboration, Jonathan has written dozens of scientific articles and a remarkable book, *Pride and a Daily Marathon*, about Ian. (He is now working on a follow-up.)[5]

5. In the early 1990s, I introduced Jonathan to my friend Marsha Ivins, an astronaut who has flown on five space shuttle missions. (She read "The Disembodied Lady," she told me, while in orbit.)

How, we wondered, might Ian do in space? The nearest thing, gravity-wise, Marsha said, was to take a ride in the astronaut training plane familiarly known as the Vomit Comet, which, by steeply climbing and then

I have found few things more moving, over the years, than to see how Jonathan, my student, has himself become eminent as a physician, physiologist, and writer; he is the author of four major books now and more than a hundred physiological papers.

After I moved to New York in 1965, I took to exploring country roads on my motorbike, looking for a suitable place to get away for occasional weekends. One Sunday, driving through the Catskills, I found a picturesque old wooden hotel perched by a lake—the Lake Jefferson Hotel. It was owned by a genial German American couple, Lou and Bertha Grupp, and we soon got to know each other. I was especially taken by their solicitude for my motorbike, which they allowed me to leave in the lobby. It soon became a familiar weekend sight to the locals. "Doc's up here again," they would say on seeing the bike.

I especially loved Saturday nights at the old bar, which was full of colorful figures, yarning and drinking, and old photographs which showed the hotel during its heyday in the 1920s and 1930s. I did much of my writing in a little alcove by the bar, where I could be alone, private, invisible, yet warmed and stimulated by the vivid life at the bar.

After a dozen or so weekends, I came to an agreement with the Grupps: I would rent a room in the basement of the hotel, come and go when I wanted, and keep my things—basically a typewriter and swimming gear—there. I could have this room

---

diving, briefly takes its passengers from almost 2 g to 0 g. Most people feel an all-encompassing weightlessness in 0 g and a corresponding heaviness in 2 g, but Ian felt neither.

and enjoy the kitchen and bar, all the amenities of the hotel, for just $200 a month.

Life at Lake Jeff was healthy and monastic. I gave up my motorcycle in the early 1970s—I had started to find the traffic in New York City too dangerous, and motorcycling was no longer a pleasure—but I always had a bicycle rack on my car, and in the long summer days I would cycle for hours. I would often stop at the old cider mill near the hotel and get two half-gallon jugs of hard cider, which I would hang on the handlebars. I love cider, and the half gallons, sipped gradually and symmetrically—a mouthful from this jug, a mouthful from that—would keep me hydrated and slightly tipsy through a long day of cycling.

There was a horse stable not far from the hotel, and sometimes I would go there on Saturday mornings and spend a couple of hours riding a giant Percheron, with a back so wide it was like bestriding an elephant. I was heavy then, over 250 pounds, but the huge animal scarcely seemed to notice my weight; it was such horses, I reflected, that carried knights and kings in full armor; Henry VIII in full armor, it was said, weighed 500 pounds.

But the greatest joy of all was swimming in the placid lake, where there might be an occasional fisherman lounging in a rowboat but no motorboats or jet skis to threaten the unwary swimmer. The Lake Jeff Hotel was past its prime, and its elaborate swimming platform and rafts and pavilions were completely deserted and quietly rotting. Swimming timelessly, without fear or fret, relaxed me and got my brain going. Thoughts and images, sometimes whole paragraphs, would start to swim through my mind, and I had to land every so often to pour them onto a yellow pad I kept on a picnic table by the side of the lake. I had such a sense of urgency sometimes that I did not have time to dry myself but rushed wet and dripping to the pad.

Eric Korn and I met in our prams, so we were told, and we have remained the closest of friends for almost eighty years. We often traveled together, and in 1979 we took a boat to Holland and rented bicycles to cycle around the country, circling back to our favorite city, Amsterdam. I had not been to Holland for some years—though Eric, living in England, had gone frequently—so I was very surprised when, completely openly, we were offered cannabis in a café. We were sitting at a table when a young man came to us and with a practiced gesture flicked open a sort of folding wallet containing a dozen or more sorts of marijuana and hash; its possession and use in modest quantities were perfectly legal in Holland by the 1970s.

Eric and I bought a packet but then forgot to smoke it. Indeed, we forgot we had it, until we got to The Hague for our ship back to England and presented ourselves at customs. We were asked the usual questions.

Had we bought anything in Holland? they asked. Liquor, perhaps?

"Yes, Genever," we replied.

Cigarettes? No, we didn't smoke.

Marijuana? Oh, yes, we had forgotten all about it. "Well, throw it away before you reach England," said the customs officer. "It's not legal there." We took it with us, thinking we might enjoy a little smoke on board.

We did have a little smoke and then threw the rest of the packet overboard. Perhaps it was more than a little smoke; neither of us had smoked for years, and the marijuana was much stronger than we expected.

I wandered off after a few minutes and found myself near the captain's wheelhouse. Illuminated in the gathering dusk, it looked enchanting, like something out of a fairy tale. The captain was navigating, his hands on the wheel, and a little boy of about ten was standing at his side, fascinated by the captain's uniform, the brass and glass dials, and the sea parting before the ship's prow. Finding the door unlocked, I entered the cabin too. Neither the captain nor the little boy by his side was disturbed by my entry, and I stationed myself quietly on the captain's other side. The captain showed us how he steered the ship, showed us all the dials; the little boy and I asked him lots of questions. We were so absorbed that we had no sense of time and were startled when the captain said that Harwich, on the English coast, was approaching. The two of us left, the little boy to find his parents, and I to find Eric.

When I found Eric, he looked haggard with anxiety, and he almost sobbed with relief as soon as he saw me. "Where were you?" he said. "I looked everywhere for you; I thought you'd jumped overboard. Thank God you're alive!" I told Eric that I had been in the captain's forecastle and enjoying myself. Then, taken aback by the intensity of his words and his expression, I said, "You care, you really care for me!"

"Of course," Eric said. "How could you doubt it?"

But it was not easy to believe that anyone cared for me; I sometimes failed to realize, I think, how much my parents cared for me. It is only now, reading the letters they wrote to me when I came to America fifty years ago, that I see how deeply they did care.

And perhaps how deeply many others have cared for me—

was the imagined lack of caring by others a projection of something deficient or inhibited in myself? I once heard a radio program devoted to the memories and thoughts of those who, like me, had been evacuated during the Second World War, separated from their families during their earliest years. The interviewer commented on how well these people had adjusted to the painful, traumatic years of their childhood. "Yes," said one man. "But I still have trouble with the three Bs: bonding, belonging, and believing." I think this is also true, to some extent, for me.

In September of 1978, I sent more of my *Leg* manuscript to Lennie—she wrote back, saying that she now felt that this might be "a happy, dancing book"—she was relieved that at last I seemed to be moving on to other interests. Towards the end of her letter, she touched on a darker matter:

> "I'm waiting to go into hospital as my very nice and good surgeon feels that the time has come for a major operation on my stupid hiatus hernia and esophagus. Pop and David don't seem very keen, but I've every confidence in him."

This was Len's last letter to me. She went into hospital, and things went wrong. What was supposed to be a straightforward operation turned into a disastrous near evisceration. When Lennie learned of this, she felt that life with intravenous nourishment and a spreading cancer was not worthwhile. She resolved to stop eating, though she would take water. My father insisted she be seen by a psychiatrist, but the psychiatrist said, "She is the sanest person I have ever seen. You must respect her decision."

I flew to England as soon as I heard about this and spent

many happy but infinitely sad days at Lennie's bedside as she was growing weaker. She was always and totally herself despite physical weakness. When I had to return to the States, I spent a morning gathering all the different tree leaves I could find on Hampstead Heath and took them to her. She loved these, identified them all, and said they took her back to her years in Delamere Forest.

I sent her a final letter at the end of 1978; I do not know if she read it:

Dearest Len,

We have all of us been hoping so intensely that this month would see your return to health; but, alas! this was not to be.

My heart is torn when I hear of your weakness, your misery—and, now, your longing to die. You, who have always loved life, and been such a source of strength and life to so many, can face death, even choose it, with serenity and courage, mixed, of course, with the grief of all passing. We, I, can much less bear the thought of losing you. You have been as dear to me as anyone in this world.

I shall hope against hope that you may weather this misery, and be restored again to the joy of full living. But if this is not to be, I must thank you—thank you, once again, and for the last time, for living—for being you.

Love,

Oliver

I am shy in ordinary social contexts; I am not able to "chat" with any ease; I have difficulty recognizing people (this is lifelong, though worse now my eyesight is impaired); I have little knowledge of and little interest in current affairs, whether political, social, or sexual. Now, additionally, I am hard of hearing, a polite term for deepening deafness. Given all this, I tend to retreat into a corner, to look invisible, to hope I am passed over. This was incapacitating in the 1960s, when I went to gay bars to meet people; I would agonize, wedged into a corner, and leave after an hour, alone, sad, but somehow relieved. But if I find someone, at a party or elsewhere, who shares some of my own (usually scientific) interests—volcanoes, jellyfish, gravitational waves, whatever—then I am immediately drawn into animated conversation (though I may still fail to recognize the person I am talking to a moment later).

I almost never speak to people in the street. But some years ago, there was a lunar eclipse, and I went outside to view it with my little 20x telescope. Everyone else on the busy sidewalk seemed oblivious to the extraordinary celestial happening above them, so I stopped people, saying, "Look! Look what's happening to the moon!" and pressing my telescope into their hands. People were taken aback at being approached in this way, but, intrigued by my manifestly innocent enthusiasm, they raised the telescope to their eyes, "wowed," and handed it back. "Hey, man, thanks for letting me look at that," or "Gee, thanks for showing me."

As I passed the parking lot opposite my building, I saw a woman arguing fiercely with the parking attendant. I went up to them and said, "Stop quarreling for a moment—look at the moon!" Startled, they stopped and looked up at the eclipse, handing the telescope to each other. Then they gave it back to me, thanked me, and instantly resumed their furious quarreling.

A similar incident happened a few years later, when I was

working on *Uncle Tungsten* and writing a chapter about spectroscopy. I had taken to wandering the streets with a tiny pocket spectroscope, peering through this at different lights and marveling at their varied spectral lines—the brilliant golden line of sodium lights, the red lines of neon, the complex lines of halogen-mercury lamps and their rare-earth phosphors. Passing by a bar in my neighborhood, I was struck by the range of colored lights within, and pressed my spectroscope against the window to examine them. It became obvious, however, that the patrons inside were disquieted by this odd behavior, my gazing at them (as they thought) with a peculiar little instrument, so I strode in boldly—it was a gay bar—and said, "Stop talking about sex, everyone! Have a look at something really interesting." There was a dumbfounded silence, but again my childish, ingenuous enthusiasm won the day, and everyone started passing the spectroscope from hand to hand, making comments like, "Wow—cool!" After everyone had had a turn with the spectroscope, it was handed back, with thanks. Then they all resumed talking about sex again.

I struggled with the *Leg* book for several more years and finally sent the completed manuscript off to Colin in January of 1983, nearly nine years after beginning it. Each section of the book, neatly typed, was on paper of a different color, though the whole manuscript was now over 300,000 words. Colin was infuriated by the sheer size of the manuscript, and its editing took virtually the whole of 1983. The final version was reduced to less than a fifth of the original size, a mere 58,000 words.

Nonetheless, it was with a sense of great relief that I relinquished the whole book to Colin. I had never been able to rid myself of a superstitious feeling that my 1974 accident was waiting to reoccur and that it would do so if I did not exorcise it by airing the entire thing in a book. Now it was done, and I was in no more danger of recapitulating the whole thing. But the unconscious is wilier than we realize, and ten days later—it was an icy day in the Bronx—I managed to fall in a particularly clumsy way and to bring about the re-accident I so feared.

I had pulled in to a gas station on City Island. I handed my credit card to the attendant, and I thought I would just open the door and stand up to stretch. The moment I got out of the car, I slipped on a patch of black ice, and when the attendant came back with the receipt, he found me on the ground, half under the car.

He said, "What are you doing?"

"Sunbathing," I replied.

And he said, "No—what happened?"

I said, "I've broken an arm and a leg," to which he replied, "You're joking again."

"No," I said, "this time I'm not joking; you'd better get an ambulance."

When I arrived at the hospital, the surgical resident asked me, "What's that written on the back of your hand?" (I had written the letters *C B S* there.)

I said, "Oh, that's a patient who has hallucinations; she's got Charles Bonnet syndrome, and I was on my way to see her."[6]

And he said, "Dr. Sacks, *you're* the patient now."

---

6. I had intended to write her story and include it in *The Man Who Mistook His Wife for a Hat,* but in the event it took more than twenty-five years to return to writing about Charles Bonnet syndrome, for *Hallucinations.*

•

When Colin heard that I was in hospital—I was still there when the proofs of the *Leg* book arrived—he said, "Oliver! You'd do anything for a footnote."

Between 1977 and 1982, *A Leg to Stand On* was finally completed, some while swimming at Lake Jeff. Jim Silberman, my editor and publisher in America, was disconcerted when I sent him the Lake Jeff section of the book. He had not received a handwritten manuscript for thirty years, he said, and this one looked as if it had been dropped in the bath. He said it would have to be not just typed but deciphered, and he sent it to one of his former editors, Kate Edgar, now freelancing in San Francisco. My illegible, water-stained manuscript with its ragged, incomplete sentences, arrows, and indecisive crossings out came back beautifully typed and annotated with wise editorial comments. I wrote to Ms. Edgar saying that I thought she had done a remarkable job with a very difficult manuscript and that she should look me up if she returned to the East Coast.

Kate returned the following year, in 1983, and she has worked with me, as editor and collaborator, ever since. I might have driven Mary-Kay and Colin mad with my many drafts, but for the past thirty years I have been lucky to have Kate working, as they did, to unmuddle, distill, and dovetail my endless drafts into a cohesive whole. (She has, moreover, been a researcher and companion on all of my subsequent books, meeting patients, listening to my stories, and sharing adventures from learning sign language to visiting chemical laboratories.)

A Matter of Identity

Although it took me almost a decade to write *A Leg to Stand On*, I was pursuing other subjects during that time as well. Chief among them was Tourette's syndrome.

In 1971, I had been re-approached by Israel Shenker, the journalist at *The New York Times* who had come to Beth Abraham in the summer of 1969 and published a long article about the initial effects of L-dopa. Now he called again to ask how the patients were doing.

Many were enjoying a sustained "awakening" with L-dopa, I replied, but some were having odd, complicated reactions to L-dopa. Above all, I said, they were ticcing. Many of them had started making sudden, convulsive movements or noises, sometimes expletives, which would burst out of them; I thought these were probably from an explosive activation of subcortical mechanisms which had been damaged by their original disease and were now being stirred up by the continual stimulation of L-dopa. I indicated to Shenker that with all of these multiple tics and expletives, some of the postencephalitics were showing something which looked like a rare condition called Gilles de la Tourette syndrome. I had never actually seen anyone with this, but I had read about it.

So Shenker came to the hospital again to observe and interview the patients. The night before the article was due to be published, I rushed to a newsstand on Allerton Avenue to get an early copy of the morning paper.

Shenker had carefully laid out the nuances of what he called "an astonishing topography of tics." He noted that one woman had an eye-clenching tic which she was able to transmute into a fist-clenching tic and that another patient could chase her tics away by concentrating on typing or knitting.

After the article appeared, I began to get a lot of letters from people with multiple tics seeking a medical opinion. I felt it would be improper to see them, because this would be profiting, in a sense, from a newspaper article. (Here, perhaps, I was echoing my father's reaction earlier that year to the *Times* review of *Migraine*.) But there was one very persistent and engaging young man whom I did see. Ray was full of convulsive tics and what he called "ticcy witticisms" and "witty ticcicisms" (he referred to himself as Witty Ticcy Ray). I was extremely fascinated by what was going on with him, not only his rapid-fire tics, but his speed of thought and wit, as well as the ways he had found to cope with his Tourette's. He had a good job and was happily married, but he could not walk down a street without everyone looking at him; he had been a target for bewildered or disapproving glances since he was five years old.

Ray sometimes thought of his Tourettic self (which he called Mr. T.) as distinct from his "real" self, much as Frances D., a normally reticent and reserved postencephalitic lady, felt that she had a "wild dopa self" very different from her civilized "real" self.

His Tourettic self made Ray impulsive and disinhibited and gave him an unusual rapidity of repartee and reaction generally. He almost always won games of Ping-Pong, not so much from skill as from the extraordinary speed and unpredictability of his serves and returns. (It had been similar in the early days when postencephalitic patients, before they were enveloped by parkinsonism and catatonia, tended to be hyperkinetic and impulsive and in this state could beat normal players in a game of football.) Ray's physiological quickness and impulsiveness, conjoined with his musicality, enabled him to be a remarkable improviser on drums.

What I had seen in the summer and fall of 1969 with the postencephalitics I thought I would never see again. Now, after meeting Ray, I realized that Tourette's syndrome was another, perhaps equally rare and rich (and in some ways kindred), subject for study.

The day after I met Ray, I thought I spotted three people in the streets of New York with the same syndrome, and the day after that another two. This amazed me, for Tourette's was described then as extremely rare, occurring in perhaps one or two people in a million. But I realized now that it must be at least a thousand times commoner. I must have been blind not to see this before, I thought, but spending time with Ray had tuned my neurological eye, so to speak, to *see* Tourette's.

I thought there must be many other people like Ray, and I fantasized about bringing them together so that they might recognize their physiological and psychological kinship and form a sort of fraternal association. In the spring of 1974, I found that this fantasy had become a reality: the Tourette Syndrome Association (TSA) had been founded in New York two years earlier by a group of parents of children with Tourette's, but it now included about twenty adults with Tourette's as well. I had seen a little girl with Tourette's in 1973, and her father, a psychiatrist who had been a founding member of the TSA, invited me to a meeting.

People with Tourette's are often unusually open to hypnosis and suggestion and disposed to involuntary repetition and imitation. I saw this at that first TSA meeting when, at one point, a pigeon flew onto a windowsill outside the conference room. It opened and closed its wings, fluttered, and then settled down. There were seven or eight people with Tourette's sitting in front of me, and I could see several of them making fluttering movements with their arms and shoulder blades, echoing the pigeon or each other.

Towards the end of 1976, at a meeting of the TSA, I was approached by John P., a young man who said, "I am the greatest Touretter in the world. I have the most complex Tourette's you will ever see. I can teach you things about Tourette's that nobody else knows. Would you like me as a specimen to study?" I was a bit taken aback by this invitation in which grandiosity and self-deprecation were strangely mixed, but I suggested we meet in my office and then decide whether further study could be productive. He presented himself to me not as someone in need of help or treatment but as a research project.

Seeing the speed and complexity of his tics and verbalizations, I thought it would be useful to have a video recorder on hand when I saw him, so I arranged to rent what at the time was the most compact recorder available—a Sony Portapak (it weighed about twenty pounds).

We had two reconnaissance sessions, and John was as good as his word. I had indeed never seen a picture as complex or as severe as the one he presented and had to live with, nor had I read or heard of anything even approaching it; I dubbed it "super-Tourette's" in my mind. I was very glad I had a video recorder going, for some of his tics and odd behaviors occurred in a fraction of a second, sometimes with two or more of them occurring simultaneously. There was far too much for the unaided eye to take in, but with the recorder I could capture everything and play it back in slow motion or frame by frame. I could also go over the video with John, who could often tell me what he had been thinking or feeling while making each tic. In this way, I thought, one might be able to have a tic analysis analogous to a dream analysis. Tics, perhaps, could be a "royal road" to the unconscious.

This was a thought I later gave up, for the majority of tics and ticcish behaviors (lunging, leaping, barking, etc.), it seemed to me, originated as reactive or spontaneous firings of the brain stem or striatum and, in this sense, were biologically but not psychically determined. But there were obvious exceptions, especially in the realm of coprolalia, the compulsive, convulsive use of expletives or offensive words (and its motor equivalent, copropraxia, or obscene gestures). John liked to attract attention, to provoke or outrage others; the compulsion to test social boundaries, the limits of propriety, is not uncommon in people with Tourette's syndrome.

I was struck in particular by a strange sound that John often uttered along with his tics. When I recorded this and slowed down the playback, elongating the sound, I discovered that it was in fact a German word—"*verboten!*"—crushed into a single unintelligible noise by its ticcish rapidity. When I mentioned this to John, he said that was how his German-speaking father would admonish him whenever he ticced as a child. I sent a copy of this tape to Luria, who was fascinated by what he called "the introjection of the father's voice as a tic."

Many tics and ticcish behaviors, I came to feel, were poised between the involuntary and the intentional, somewhere between jerks and acts, subcortical in origin but sometimes given meaning and intentionality whether conscious or subconscious.

One summer day, while John was in my office, a butterfly flew in through the open window. John followed its soaring, zigzag flight with sudden, erratic jerks of his head and eyes as he poured out a stream of endearments and imprecations: "I want to kiss you, I want to kill you," he repeated, and then abbreviated this to "kiss you, kill you, kiss you, kill you." After two or three minutes of this—he seemed unable to stop as long as the butterfly was fluttering around—I said, jokingly, "If you were really concentrating, you could disregard the butterfly, even if it landed on your nose."

The moment I said this, he clutched the end of his nose and tore at it, as if to dislodge an enormous butterfly which had attached itself there. I wondered if his excessively vivid Tourettic imagination had crossed over into hallucination, conjuring up a phantom butterfly as real, perceptually, as an actual one. It was like a little nightmare being enacted in full consciousness before me.

I worked intensively with John in the first three months of 1977, and this produced a sense of wonder, discovery, and intellectual excitement more intense than anything I had felt since the summer of 1969, when the postencephalitics were awakening. It revived in me very strongly the sense which I had after meeting Ray, that I must write a book on Tourette's. I wondered about writing a book with John as a central character—perhaps a composite or actual "day in the life" of someone with super-Tourette's.

After so promising a start, I thought that a full-scale study could be immensely informative, but I cautioned John, saying that such a study was essentially an exploration, an investigation, and no therapeutic benefit could be promised. In this way, the study might be akin to Luria's *Mind of a Mne-*

monist or to Freud's *Interpretation of Dreams* (I had these two books constantly by my side during the months of our "Tourette-analysis").

I would see John in my office every Saturday, and I recorded our sessions with two video recorders going simultaneously, one narrowly focused on John's face and hands, the other with a wider-angle view of the two of us.

Driving to my office on Saturday mornings, John would often stop at an Italian grocery on the way and get himself a sandwich and a Coke. The grocery was popular, always full of people whom John could describe, or rather impersonate, uncannily, bringing them all to life. I had been reading Balzac and quoted him to John as saying, "I have a whole society in my head."

"So do I," said John, "but in the form of imitation." These instant, involuntary imitations and mimicries often had a flavor of caricature or mockery, and John would sometimes draw amazed or outraged glances from those around him, which he in turn impersonated or caricatured. Sitting in my office, hearing him describe and enact such scenes, I started to think that I might need to go out into the world with him to witness such interactions for myself.[1] I felt very hesitant about doing this; I did not want to make him self-conscious, feeling he was observed all the while (or literally "on camera," if I had the Portapak with me), nor did I want to intrude too much into his life outside our Saturday morning routine. And yet I thought it would be of great value if one could record a day or a week in the life of such a super-Touretter—that this could provide an

---

1. I described the first time I did so in "The Possessed," a chapter of *The Man Who Mistook His Wife for a Hat,* though I disguised John P.'s identity by portraying him as an old woman.

anthropological or ethological view to complement the clinical and phenomenological observations made in the office.

I contacted a team of anthropological documentary filmmakers—they were just back from filming a tribe in New Guinea—and they were intrigued by the notion of a sort of medical anthropology. But they wanted $50,000 for a week's recording, and I did not have $50,000; it was more than I earned in an entire year.

I mentioned this to Duncan Dallas (I knew that Yorkshire Television sometimes gave grants for documentary fieldwork), and he said, "Why don't I come and see him?" Duncan arrived a couple of weeks later, and he agreed that John was unlike anything he had ever seen before and that he was very articulate and good at presenting himself. Duncan wanted to make a full documentary about him, and John, who had seen the *Awakenings* documentary, was excited by the idea. By this time, however, I was less enthusiastic and a little disturbed by what seemed to be excessive enthusiasm and perhaps expectations on John's part. I wanted to continue my quiet, exploratory work with him, and now he was dreaming of being the central figure in a television documentary.

He had said that he liked to "perform," to make "scenes," to be the center of attention, but would afterwards avoid ever going back to a place where he had once created such a scene. How might he react to having some of his "scenes" or "performances"—exhibitionistic, but arising from his tics—captured on film, given a permanent form which he could not erase? All this was carefully discussed by the three of us during Duncan's reconnaissance visit, and Duncan was at pains to say that John could come over to England and play a part in the editing of the film at any stage.

The filming was done in the summer of 1977, and John was in his best form: full of tics and antics, driven but also playful—clowning, improvising, and imitating when he had an audience, but also talking carefully, soberly, and often very movingly about life for someone like himself. We all thought that a remarkable but balanced and very human documentary would result.

After the filming, John and I went back to our own quiet sessions together, but I observed a certain tension in him now—a holding back I had not seen before—and when he was invited to go to London to take an active part in the editing, he declined the invitation.

The film was shown on British television early in 1978; it attracted much notice, all of it good, and a slew of letters were sent to John by viewers who felt for him and admired him. He was at first very proud of the documentary and showed it to his friends and neighbors, but he then became deeply disturbed and, above all, angry about it and turned on me, saying that I had "sold" him into the hands of the media (forgetting that he was the one who had most wanted a film and that I was the one who had counseled caution). He wanted the film to be suppressed and never shown, and this equally for the videotapes I had made (which now numbered over a hundred). If the film were ever shown again, if ever any of the tapes were shown, he said, he would come after me and kill me. I was deeply shocked and bewildered by all this—scared, too—but I acceded to his wishes, and the documentary has never been broadcast again.

But this, alas, did not satisfy him. He started to make threatening phone calls to me, and these, at first, consisted of two words, "Remember Tourette," because as he knew I knew perfectly well, Gilles de la Tourette had himself been shot in the head by one of his own patients.[2]

Under the circumstances, I could not show any of the footage of John even to my medical colleagues, and this was intensely frustrating, because I felt it was extremely valuable material that could cast illumination not just on many aspects of Tourette's syndrome but on rarely explored aspects of neuroscience and human nature in general. I thought I could write an entire book based on five seconds of this videotape, but I never did.

I withdrew an article I had written about John for *The New York Review of Books*; it was already in proof, but to publish, I now feared, would have inflamed him.

I understood more when the documentary of *Awakenings* was shown at a psychiatric meeting in the fall of 1977, and its screening was continually interrupted by someone who turned out to be John's sister. We spoke later, and she said she found the documentary and its exposure of such patients "shocking." She was alarmed that her brother would be exposed on television; people like him, she added, should be hidden from sight. I began to appreciate, too late, the depths of John's ambivalence about filming: his compulsion to be seen and shown, to exhibit himself, but also to hide from sight.

In 1980, as a break from my abortive struggles with the *Leg* book, I wrote a piece about Ray, the charming, witty, ticcy

---

2. Gilles de la Tourette's patient in fact did not have Tourette's syndrome but an erotic fixation on him; such fixations, as one saw in the case of John Lennon, can lead to murder. Tourette himself was rendered hemiplegic and aphasic by the bullet wound.

man whom I had seen and followed for almost ten years. I was anxious about Ray's reaction to my writing about him, so I asked how he might feel about its being published and offered to read my piece to him.

He said, "No, that's okay. You don't have to."

When I insisted, he invited me to his house for dinner so I could read it to him and his wife afterwards. Ray ticced and twitched a good deal as I read, and at one point he burst out, "You take a few liberties!"

I stopped and pulled out a red pencil, saying, "What shall I delete? It's up to you."

But he said, "Go on—keep reading."

When I came to the end of the piece, he said, "It's essentially true. But don't publish it here. Publish it in London."

I sent the piece to Jonathan Miller, who liked it and passed it on to Mary-Kay Wilmers, who (with Karl Miller, Jonathan's brother-in-law) had recently founded the *London Review of Books*.

"Witty Ticcy Ray" was a different sort of writing from anything I had done before—it was the first full-length case history I had written about living a life, a full life, despite a complex neurological condition—and its reception encouraged me to write more case histories like it.

•

In 1983, Elkhonon Goldberg, a friend and colleague who had studied in Moscow under Luria, asked me if I would join him in giving a seminar at the Albert Einstein College of Medicine on the new field of neuropsychology which Luria had pioneered.

This session was devoted to agnosias—perceptions or misperceptions stripped of meaning—and at one point Goldberg turned to me and asked if I could give an example of a visual agnosia. I immediately thought of one of my patients, a music teacher who had become unable to recognize his students (or anyone else) visually. I described how Dr. P. might pat the "heads" of water hydrants or parking meters, mistaking them for children, or amiably address knobs on the furniture and be astounded when they did not reply. At a certain point, I said, he even mistook his wife's head for a hat. The students, while appreciating its gravity, could not help laughing at this comical situation.

I had not thought of elaborating my notes on Dr. P. up to this point, but telling his story to the students brought our encounter back into my mind, and that evening I wrote up his case history. I titled it "The Man Who Mistook His Wife for a Hat" and sent it off to the *London Review of Books*.

It did not occur to me that it might become the title story of a collection of case histories.

•

In the summer of 1983, I went to Blue Mountain Center, a retreat for artists and writers, for a month. It was on a lake, which was wonderful for swimming, and I had my mountain bike with me. I had never found myself surrounded by writers and artists before, and I enjoyed the combination of solitary days spent writing and thinking and convivial dinners with the other residents at the end of the day.

For the first two weeks at Blue Mountain, however, I found myself completely blocked and in great pain: I had pushed myself too hard on the bike, and my back had seized up. On day sixteen, I happened to pull out the memoirs of Luis Buñuel to read and found a sentence in which he expresses his fear of losing his memory and his identity, as his aging and demented mother had done. This suddenly activated my memories of

Jimmie, an amnesic sailor I had started seeing in the 1970s. I got to work instantly, spent twelve hours writing about Jimmie, and by nightfall completed his story, "The Lost Mariner." I wrote nothing more from days seventeen to thirty. When people asked me if my stay at Blue Mountain had been a "productive" one, I was not sure how to answer: I had one supremely productive day and twenty-nine blocked or sterile ones.

I offered the piece to Bob Silvers at *The New York Review of Books*, and he liked it, although he made an interesting request. He asked, "Can I see your notes on the patient?" He looked through my consultation notes, written each time I had seen Jimmie, and he said, "Many of these are more vivid and immediate than what you've given me. Why don't you insert some of your notes and weave things together, so we have both your immediate response to the patient and the more reflective one as you look back over the years?"[3] I followed his advice, and he published the piece in February of 1984. This was a huge encouragement to me, and over the next eighteen months I sent him five more pieces, which would form a nucleus for *The Man Who Mistook His Wife for a Hat*. Bob's support and friendship and his endlessly careful and constructive editing are legendary; once he phoned me when I was in Australia, asking how I would feel about replacing a comma with a semicolon. And he has nudged me into many essays I might not have written otherwise.

I continued to publish individual pieces (some in *The New*

3. When "The Lost Mariner" came out, it elicited a letter from Norman Geschwind, one of the most original and creative neurologists in America. I was very excited at hearing from him. I wrote back immediately, but there was no reply, because Geschwind had just had a catastrophic stroke. He was only fifty-eight, but he left an enormous legacy.

York Review of Books, others in different journals, like *The Sciences* and *Granta*), at first with no notion that they could come together in any way. Colin and Jim Silberman, my American publisher, felt there was a sort of unity of tone and feeling in them, but I was not sure they would hold together as a book.

I wrote what were to become the last four pieces in *The Man Who Mistook His Wife for a Hat* on the last four days of 1984, conceiving them as a quartet, perhaps even a little book to be called "The World of the Simple."

The following month I visited my friend Jonathan Mueller, who was working as a neurologist at the VA Hospital in San Francisco. As we walked around the Presidio, where the hospital was located, he told me of his interest in the sense of smell. I then told him two stories. One was about a man who, despite complete and permanent destruction of his sense of smell from a head injury, started to imagine (or possibly hallucinate) contextually appropriate smells, like the smell of coffee when he saw it brewing. The other was the story of a medical student who, in the course of an amphetamine-induced mania, developed an extraordinary heightening of the sense of smell (this story was really about my own experience, though in *Hat* I called the medical student "Stephen D."). The next morning, over a very long meal in a Vietnamese restaurant, I wrote both stories out, bracketed them under a single title ("The Dog Beneath the Skin"), and sent them off to my publishers. I had felt that my *Hat* book was missing something, and now "Dog" provided the missing piece.

With this, I felt a marvelous sense of completion and liberation. I had completed my book of "clinical tales," I was a free man, and I could take a real holiday, which I felt I had not done in a dozen years. Impulsively, I decided to visit Australia;

I had never been there, and my brother Marcus lived in Sydney with his wife and children. I had met Marcus's family when they came to England for my parents' golden wedding anniversary in 1972 but had not seen them since. I walked down to San Francisco's Union Square, where Qantas had an office, presented my passport, and said I wanted to go on the first available flight to Sydney. No problem, they said, there was plenty of room, and I had just enough time to hurry back to the hotel, get my things, and head for the airport.

It was the longest plane ride I had ever been on, but time passed rapidly as I wrote excitedly in my journal, and fourteen hours later we arrived in Sydney; I recognized the famous bridge and the opera house as we circled the city. I handed in my passport at Passport Control and was about to move on when the passport officer said, "And your visa?"

"Visa?" I replied. "What visa? Nobody told me about a visa." Previously genial, the passport officer suddenly became very stern and serious: Why was I coming to Australia? Was there anyone who could vouch for me? I said that my brother and his family were in the airport awaiting me. I was told to sit down while he was found and my bona fides checked. The authorities gave me a provisional visa for ten days, but they warned me, "Never do this again, or we'll send you straight back to the States."

My ten days in Australia gave me a strong sense of joyous discovery—discovering a brother I hardly knew (Marcus was ten years older than I and had gone to Australia in 1950); a sister-in-law, Gay, whom I immediately felt at home with (she shared my passions for minerals and plants, for swimming and diving); and a young nephew and niece who bonded with their new (and, to their eyes, exotic) uncle.

With Marcus, I sought and found a relationship I had never really had with my brothers in England. It was a relationship I could not have had with David, who was so unlike me—dapper, charming, social—or Michael, who was lost in the depths of schizophrenia. With Marcus—quiet, scholarly, thoughtful, and warm—I felt I could have a deeper sort of relationship.

I fell in love, too, with Sydney and later with the Daintree Rainforest and the Great Barrier Reef in Queensland, which I found overwhelmingly beautiful—and strange. Seeing Australia's unique flora and fauna made me think of how Darwin was so astonished by the plants and animals in Australia that he wrote in his journal, "Two distinct Creators must have been at work."

·

After the heights and depths Colin and I had been through with *Awakenings* and *Leg,* our relationship became lighter and easier. If the yearlong editing of *Leg* had almost killed us both, working on the *Hat* book, as we both called it, was straightforward. Many of the pieces in *Hat* had already been published, and Colin, besides editing the rest, suggested how they might be divided into four groups, with introductions for each section.

Colin published the book in November 1985, just six months after the manuscript was completed; the U.S. edition came out in January 1986, with a modest initial printing of fifteen thousand copies.

My *Leg* book had not sold particularly well, and no one expected that a book of neurological stories would be a commercial success. But after a few weeks, Summit had to do another printing and then another. The book's popularity grew through word of mouth, and in April, completely unexpectedly, it appeared on the *New York Times* best-seller list. I thought it must be a mistake or a temporary surge, but it would stay on the best-seller list for twenty-six weeks.

What amazed and moved me, even more than being a "best seller," were the letters which poured in, many from people who had themselves experienced problems which I had written about in *Hat*—face-blindness, musical hallucinations, etc.— but which they had never before admitted to anyone or even, sometimes, to themselves. Others asked about the people I had written about.

"How is Jimmie, the Lost Mariner?" they would write. "Say hello to him. Give him my best." Jimmie was real to them, and this was so of many other figures in the book; the reality of their situations and struggles touched the hearts, as well as the minds, of many readers. Readers could imagine themselves in Jimmie's position, whereas the extreme and tragic predicament of my *Awakenings* patients was almost beyond even the most sympathetic imagination.

One or two reviewers saw me as specializing in the "bizarre" or "exotic," but I felt the opposite. I thought of my case histories as "exemplary"—I very much liked Wittgenstein's dictum that a book should consist of examples—and hoped that perhaps, by depicting cases of exceptional severity, one might illuminate not only the impact and experience of being neurologically ill but crucial and perhaps unexpected aspects of the organization and workings of the brain.

•

Although Jonathan Miller had said to me, after *Awakenings* was published, "You're famous now," it was not really so. *Awakenings* had received a literary prize and acclaim in England, but it was hardly noticed in the United States (it got only one review, from Peter Prescott at *Newsweek*). With the sudden popularity of *Hat*, though, I had entered the public sphere, whether I wanted it or not.

Certainly there were advantages. Suddenly I was in contact with a great many people. I had powers to help but also powers to harm. I could no longer write anything in an anonymous way. I had not really thought of a reading public when I wrote *Migraine* and *Awakenings* and the *Leg* book. Now I felt a certain self-consciousness.

I had given occasional public lectures before, but after *Hat* came out, I was deluged with speaking invitations and requests of all sorts. For better and for worse, with *Hat*'s publication, I became a public figure, with a public persona, even though by disposition I am solitary and venture to believe that the best, at least the most creative, part of me is solitary. Solitude, creative solitude, was more difficult to come by now.

My fellow neurologists, however, remained somewhat remote and dismissive. Now to this was added, I think, a certain suspicion. I had, it seemed, defined myself as a "popular" writer, and if one is popular, then, ipso facto, one is not to be taken seriously. This was by no means completely so, and there were some colleagues who saw *Hat* as solid, detailed neurology embedded in a fine, classical narrative form. But by and large, the medical silence continued.

In July of 1985, a few months before *Hat* was published, I felt a rebirth of interest in Tourette's syndrome. Over the course of a few days, I filled an entire notebook with thoughts and again saw the possibility of a whole book arising. I was visiting England at the time, and this flow of ideas and excitement reached its height on the plane back to New York. But it was to be interrupted a day or two after my return, when the postman delivered a package to my little house on City Island. It was from *The New York Review of Books,* and it contained Harlan Lane's history of the deaf and of sign language, *When the Mind Hears.* Bob Silvers wanted to know if I would review it. "You have never really thought about language," Bob wrote. "This book will force you to."

I was not sure I wanted to be diverted from the Tourette's book I planned to write. I had originally wanted to start a Tourette's book after meeting Ray in 1971, but it was derailed first by my leg accident and then by the business with John. Now I felt it was in danger of being bumped once again. And yet Harlan Lane's book both fascinated and outraged me. It recounted the story of deaf people; their unique, rich culture based on a visual language, sign; and the continuing debate about whether deaf people should be educated in their own visual language or forced into "oralism," an often disastrous decision for people born deaf.

My interests in the past had always risen directly from clinical experience, but now I found myself, almost against my will, becoming involved in an exploration of deaf history and culture and the nature of sign languages—something I had no firsthand experience with. But I went to visit some local deaf schools where I met a number of deaf children. And, inspired by Nora Ellen Groce's book *Everyone Here Spoke Sign Language,* I visited a small town in Martha's Vineyard where a century before nearly a quarter of the population had been born deaf. Deaf people in this town were not seen as "deaf"; they were seen simply as farmers, scholars, teachers, sisters, brothers, uncles, aunts.

By 1985, there were no longer any deaf people in the town, but the older hearing people still had vivid memories of their deaf relatives and neighbors and still sometimes used sign among themselves. Over the years, the community had adopted the one language everyone could use; hearing and deaf alike were fluent in sign. I had never really thought much about cultural topics, and I was intrigued to encounter the idea of an entire community adapting in this way.

When I visited Gallaudet University in Washington, D.C., (it is the only university in the world for deaf and hearing-impaired students) and talked about the "hearing impaired," one of the deaf students signed, "Why don't you look at yourself as sign impaired?"[4] It was a very interesting turning of the tables, because there were hundreds of students all conversing in sign, and I was the mute one who could understand nothing and communicate nothing, except through an interpreter. I got drawn more and more deeply into deaf culture, and my little book review expanded into a more personal essay, which was published in *The New York Review of Books* in the spring of 1986.

And that, I thought, was the end of my involvement with the deaf world—a brief but fascinating journey.

One day in the summer of 1986, I got a phone call from a young photographer, Lowell Handler. He had been using special stroboscopic techniques to catch people with Tourette's syndrome in mid-tic. Could he come round and show me his

---

4. I longed to be able to communicate with deaf people in their own language, and Kate and I took American Sign Language classes for many months, but, alas, I am terrible at learning other languages and was never able to sign more than a few words and phrases.

portfolio? He had a special sympathy for the subject, he said, because he himself had Tourette's. A week later we met. His portraits impressed me, and we began to discuss a possible collaboration in which we would travel around the country meeting other people with Tourette's, documenting their lives in text and photographs.

Both of us had heard tantalizing reports about a Mennonite community in a small town in Alberta where there was an extraordinary concentration of people with Tourette's syndrome. Roger Kurlan and Peter Como, neurologists in Rochester, had visited La Crete several times to map the genetic distribution of Tourette's, and some in the Tourette's community had jokingly started to refer to the town as Tourettesville. But there had not been a detailed study of particular individuals in La Crete and of what it meant to have Tourette's syndrome in such a close-knit, religious community.

Lowell went on a preliminary visit to La Crete, and we started to plan a longer expedition. We needed support for the cost of travel and processing a great deal of film. I applied to the Guggenheim Foundation for a fellowship, proposing to do a study of the "neuroanthropology" of Tourette's syndrome, and was awarded a grant for $30,000; Lowell got a commission from *Life* magazine, still flourishing at the time and famous for its photojournalism.

By the summer of 1987, arrangements had been made for us to visit La Crete. Lowell was laden with cameras and extra lenses; I had just my usual notebooks and pens. The La Crete visit was extraordinary in many ways and broadened my sense of the range of Tourette's syndrome and of people's reactions to it. It also gave me a sense of how strongly Tourette's, though neurological in origin, could be modified by context and culture—in this case, a deeply supportive religious community in which Tourette's was accepted as God's will. What would living with Tourette's be like, we wondered, in a much more permissive environment? We resolved to visit Amsterdam to find out.[5]

Lowell and I stopped in London on our way to Amsterdam, in part because I wanted to visit my father on his birthday (he was ninety-two) and in part because *Hat* had just come out in paperback and the BBC had asked me to talk about Tourette's on its World Service. There was a taxi waiting to take me back to the hotel after my interview, with a most unusual taxi driver. He jerked, ticced, barked, cursed, and at a red light got out and jumped on the hood of his car, leaping back into his driver's seat before the light turned green. I was amazed at this—how clever of the BBC or my publishers, knowing I would be talking about Tourette's, to select a spectacularly Tourettic taxi driver to drive me back! And yet I was puzzled. The taxi driver did not say anything, although he must have known that my special interest in Tourette's was the reason he had been selected. I said nothing for several minutes and then, hesitantly, asked him how long he had had the condition.

"What do you mean—'condition'?" he said angrily. "I don't have a 'condition'!"

I apologized and said I had not meant to upset him, but I was a doctor and had been so struck by his unusual movements that I wondered if he had a condition called Tourette's syndrome. He shook his head violently and repeated that he did not have any "condition" and that if he had a few nervous

---

5. In a forthcoming book, I describe in more detail our travels in Canada and Europe and across the United States.

movements, this had not prevented him from being a sergeant in the army or anything else. I said nothing more, but when we arrived at my hotel, the driver said, "What was that syndrome you mentioned?"

"Tourette's syndrome," I said, and gave him the name of a neurological colleague in London, adding that she was a very warm, understanding person, as well as having an unrivaled expertise with patients with Tourette's.

•

The Tourette Syndrome Association had grown steadily since 1972, and satellite groups were being established all over the United States (and indeed around the world). In 1988, the TSA organized its first national meeting, and nearly two hundred people with Tourette's gathered for three days at a hotel in Cincinnati. Many of them had never met another person with Tourette's, and they feared they might "catch" one another's tics. This fear was not unfounded, for sharing tics can indeed occur when people with Tourette's meet. Indeed, some years ago, after meeting a Tourettic man in London with a spitting tic, I mentioned this to another Touretter in Scotland, who immediately spat and said, "I wish you hadn't told me that!" and added a spitting tic to his already large repertoire.

In honor of the Cincinnati meeting, the governor of Ohio had declared a statewide Tourette's awareness week, but apparently not everyone knew this. One of us, Steve B., a young man with striking Tourette's and coprolalia, went into a Wendy's restaurant for a hamburger. While he was waiting for his food, Steve jerked and yelled an obscenity or two, and the restaurant manager asked him to leave, saying, "You can't do that here."

Steve said, "I can't help it; I have Tourette's syndrome." He showed the manager an informational pamphlet from the

TSA meeting, adding, "This is Tourette's Syndrome Awareness Week—haven't you heard?"

The manager said, "I don't care; I've already called the police. Leave now, or you will be arrested."

Steve returned to the hotel, outraged, and told us the story, and pretty soon there were two hundred Touretters marching on Wendy's, ticcing and shouting, and I was in their midst. We had alerted the media, the Ohio press took up the story, and I suspect Wendy's has never been the same. This is the only time in my life I have been involved in any sort of demonstration or march, apart from one other occasion, also in 1988.

In March of 1988, quite unexpectedly, Bob Silvers called. "Have you heard about the revolution of the deaf?" he asked. There had been a revolt of the deaf students at Gallaudet, who were protesting the appointment of a hearing president for the university. They wanted a deaf president, a president who could communicate in fluent American Sign Language, and they had barricaded the campus, shutting the school down. I had been to Gallaudet a couple of times, Bob said—would I go back to Washington to cover the revolt? I agreed to go, and I invited Lowell to come and take photographs. I asked my friend Bob Johnson, a professor of linguistics at Gallaudet, to be our interpreter.

The Deaf President Now protest lasted more than a week, culminating in a march on the Capitol (Gallaudet was founded and maintained by a congressional charter). My role as impartial observer soon got compromised; I was walking along and making notes on the sidelines when one of the deaf students grabbed me by the arm and signed, "Come on, you're with us."

So I joined the students—over two thousand of them—in their protest march. The essay I wrote about this for *The New York Review of Books* was the first "reporting" I had ever done.

Stan Holwitz at the University of California Press (which had published *Migraine* in the United States) suggested that my two essays on the deaf might make a nice book, and I liked the idea but felt I needed to write a few paragraphs as a sort of bridge between the two parts—something on general aspects of language and the nervous system. I had no inkling that these few paragraphs would in fact become the largest part of the book, which would eventually be called *Seeing Voices.*

A Leg to Stand On had got many nice reviews when it came out in England in May of 1984, but these had been eclipsed in my mind by a single highly critical one, by the poet James Fenton. His review upset me deeply and brought me to a depressive halt for three months.

But when the American edition came out later in the year, I was elated by a wonderful, generous review in *The New York Review of Books*, a review which so revived and energized and reassured me that it led to an explosion of writing—twelve pieces in a few weeks, completing *The Man Who Mistook His Wife for a Hat.*

The review was by Jerome Bruner, who was a legendary figure, a founder of the cognitive revolution in psychology in the 1950s. At that time, behaviorism, as espoused by B. F. Skinner and others, was dominant; one looked only at stimulus and response—the visible, public manifestations of behavior. There was no reference to any inner process, to what might go on internally *between* stimulus and response. The concept of "mind" scarcely existed for Skinner, but this was exactly what Bruner and his colleagues set out to restore.

Bruner was a close friend of Luria's, and they had many intellectual affinities. In his autobiography, *In Search of Mind*, Bruner described meeting Luria in Russia in the 1950s. He wrote, "Luria's [view] on the role of language in early development was more than welcome to me. So were his other enthusiasms."

Like Luria, Bruner insisted on observing children as they acquired language not in a lab setting but in their own environment. In *Child's Talk*, he greatly enlarged and enriched our concepts of how language is acquired.

In the 1960s, following Noam Chomsky's revolutionary work, there was a great emphasis on syntax in linguistics; Chomsky postulated that the brain had a built-in "language acquisition device." This Chomskyan sense of a brain hardwired and poised to acquire language on its own seemed to ignore its social origins and its fundamental function as communication. Bruner argued that grammar could not be dissociated from meaning or communicative intent. In his view, the syntax, the semantics, and the pragmatics of language all went together.

It was Bruner's work, above all, which allowed me to think about language not just in linguistic terms but in social ones as well, and this was crucial for my understanding of sign language and deaf culture.

Jerry has been a good friend and, at one level and another, a sort of guide and implicit mentor. There seem no limits to his curiosity and knowledge. He has one of the most spacious, thoughtful minds I have ever encountered, with a vast base of knowledge of every sort, but it is a base under continual questioning and scrutiny. (I have seen him suddenly stop in mid-sentence and say, "I no longer believe what I was about to say.") At ninety-nine, his remarkable powers seem undiminished.

•

Although I had observed *losses* of language—various forms of aphasia—in my patients, I was very ignorant about the development of language in children. Darwin had portrayed the development of language and mind in his lovely "Biographical Sketch of an Infant" (the infant being his firstborn son), but I had no children of my own to observe, and none of us have any personal memory of that crucial period when language is acquired in the second or third year of life. I needed to find out more.

One of my closest friends at Einstein was Isabelle Rapin, a pediatric neurologist from Switzerland who was especially interested in the neurodegenerative and neurodevelopmental disorders of childhood. This was one of my own interests at the time; I had written a paper on "spongy degeneration" (Canavan's sclerosis) in identical twins.

The neuropathology department organized brain cuttings once a week, and it was at one of these, soon after I started at Einstein, that I met Isabelle.[6] We formed an unexpected pair—Isabelle, precise and rigorous in her thinking, and me, slovenly,

---

6. These brain cuttings were popular sessions which attracted, among others, clinicians eager to see if their diagnoses held up. On one memorable occasion, we examined the brains of five patients who had been diagnosed in life as having multiple sclerosis. The brain cuttings, however, revealed that all of them had been misdiagnosed.

slapdash, full of odd associations and mental detours—but we hit it off from the start, and we have remained very close friends.

Isabelle would never permit me, any more than she permitted herself, any loose, exaggerated, or uncorroborated statements. "Give me the evidence," she always says. In this way, she is my scientific conscience, and she has saved me from many embarrassing blunders. But when she feels that I am on solid ground, she insists that I publish what I have observed plainly and clearly, so it can be properly seen and debated, and in this way she has been behind many of my books and articles.

I would often ride my motorbike up to Isabelle's weekend house on the banks of the Hudson, and Isabelle, Harold, and their four children made me a part of their family. I would come up and spend the weekend talking with Isabelle and Harold, occasionally taking the children for a motorbike ride or a swim in the river. In the summer of 1977, I lived in their barn for a whole month, working on an obituary of Luria.[7]

A few years later, when I began thinking and reading about deafness and sign language, I spent an intense three-day weekend with Isabelle, and she spent many hours educating me about sign language and the special culture of the deaf, which she had observed over many years working with deaf children.

She drummed into me what Luria's mentor, Vygotsky, had written:

---

7. In a September 1977 letter, Auntie Len thanked me for my birthday telegram ("[It] warmed the cockles of my eighty-five-year-old heart") but went on to say, "We were shocked that Professor Luria died. It must be a great personal blow for you. I know how much you valued his friendship. Did you write the *Times* obituary?" (I did.)

If a blind or deaf child achieves the same level of development as a normal child, then the child with a defect achieves this *in another way, by another course, by other means.* And, for the pedagogue, it is particularly important to know the *uniqueness* of the course, along which he must lead the child. The key to originality transforms the minus of the handicap into the plus of compensation.

The monumental achievement of learning language is relatively straightforward—almost automatic—for hearing children but may be highly problematic for deaf children, especially if they are not exposed to a visual language.

Deaf, signing parents will "babble" to their infants in sign, just as hearing parents do orally; this is how the child learns language, in a dialogic fashion. The infant's brain is especially attuned to learning language in the first three or four years, whether this is an oral language or a signed one. But if a child learns no language at all during the critical period, language acquisition may be extremely difficult later. Thus a deaf child of deaf parents will grow up "speaking" sign, but a deaf child of hearing parents often grows up with no real language at all, unless he is exposed early to a signing community.

For many of the children I saw with Isabelle at a school for the deaf in the Bronx, learning lip-reading and spoken language had demanded a huge cognitive effort, a labor of many years; even then, their language comprehension and use was often far below normal. I saw how disastrous the cognitive and social effects of not achieving competent, fluent language could be (Isabelle had published a detailed study of this).

With my interest in perceptual systems in particular, I wondered what went on in the brain of a person born deaf, especially if his native language was a visual one. I learned that very recent studies had shown how, in the brains of congenitally deaf signers, what would normally be auditory cortex in a hearing person was "reallocated" for visual tasks and especially for the processing of a visual language. The deaf tend to be "hypervisual" compared with hearing people (this is evident even in the first year of life), but they become much more so with the acquisition of sign.

The traditional view of the cerebral cortex was that each part of it was pre-dedicated to particular sensory or other functions. The idea that parts of the cortex could be reallocated to other functions suggested that the cortex might be much more plastic, much less programmed, than formerly thought. It made clear, through the special case of the deaf, that an individual's experience shapes the higher functions of his brain by selecting (and augmenting) the neural structures which underlie function.

This seemed to me of momentous importance, something which demanded a radically new vision of the brain.

City Island

Though I had left the West Coast for New York in 1965, I stayed in close touch with Thom Gunn, visiting him whenever I was in San Francisco. He now shared an old house with Mike Kitay and, as far as I could judge, four or five other people. There were thousands of books, of course—Thom's reading was serious, incessant, passionate—but there was also a collection of beer advertisements going back to the 1880s, records galore, and a kitchen full of intriguing spices and smells. Thom and Mike both enjoyed cooking, and the house itself had a sweet flavor, alive with personalities and idiosyncrasies, people wandering to and fro. Having always been solitary myself, I enjoyed these brief glimpses of communal living, which seemed to me full of affection and accommodation (no doubt there were conflicts, too, but I was largely unaware of them).

Thom was always a tremendous walker, striding up and down the hills of San Francisco. I never saw him with a car or a bicycle; he was quintessentially a walker, a walker like Dickens, who observed everything, took it in, and used it, sooner or later, in what he wrote. He liked to prowl around New York, too, and when he visited, we would take a ride on the Staten Island Ferry, or a train to some out-of-the-way place, or just go for a walk around the city. We would usually end up in a restaurant, though once I tried to make a meal at my place. (Thom was on antihistamines at the time and felt too sedated to go out.) I am no cook, and everything went wrong; the curry blew up, and I got covered with yellow powder. This incident must have stuck in his mind, for when he sent me his poem "Yellow Pitcher Plant" in 1984, he inscribed the manuscript "For saffron-handed Sacks from Dozy Gunn."[1]

In an accompanying letter, he wrote,

> How good to see you, o saffron-handed one! I may have seemed in a doze of antihistamines, but I was attentive and interested at the center. I have thought about what you said of anecdote and narrative. I think we all live in a swirl of anecdotes. . . . We (most of us) compose our lives into narratives. . . . I wonder what the origin is of the urge to "compose" oneself.

•

We never knew where the conversation would go. On that day, I had read Thom part of an as-yet-unpublished piece about Mr. Thompson, an amnesic patient who had to make himself and his world up every moment. Each of us, I had written, constructs and lives a "narrative" and is defined by this narrative. Thom was fascinated by stories about patients and would often draw me out on the subject (though I needed little encouragement to get going). Looking through our correspondence, I find, in one of his first letters to me, "It was good seeing you last weekend, and Mike and I have been thinking about phantom limbs ever since," and in another letter, "I remember your discourse about Pain. That will be a fine book too." (Alas, it was never written.)

---

1. Knowing of my botanical side, Thom would send me all of his "plant" poems. After getting his "Nasturtium," I wrote, "I hope you may write more poems like this, celebrations of brave plants in vacant lots, ditches, crannies, etc.—you remember how the figure of Hadji Murad came back to Tolstoy when he saw a crushed but still fighting thistle by the roadside."

Though Thom started in the 1960s to send me all his books (always with charming and idiosyncratic inscriptions), it was only after *Migraine* came out early in 1971 that I was able to reciprocate. After this the stream ran in both directions, and we wrote to each other regularly (my letters often ran to several pages, while his, incisive and to the point, often came on postcards). We occasionally talked about the process of writing, the rushes and stoppages, the illuminations and darknesses, which seemed to be part and parcel of the creative process.

I had mentioned to him, in 1982, that the almost unbearable delays and breaks and losses of enthusiasm in the composition of my *Leg* book seemed finally to be coming to an end after eight years. Thom wrote back,

> I always felt thwarted you denied us *A Leg to Stand On,* though perhaps that might yet reach us in a revised version. . . . I am a bit slothful at the moment. My pattern seems to be: a long cessation of any coherent writing after I have completed a MS, then a tentative start followed by, during the next few years, various separate bursts of activity, ending with a sense of the new book as a whole, in which I make discoveries about my subject(s) that I have never anticipated. It's strange, the psychology of being a writer. But I suppose it's better not to be merely facile—the blocks, the feelings of paralysis, the time when language itself seems dead, these all help me in the end, I think, because when the "quickenings" do come they are all the more energetic by contrast.

It was crucial for Thom that his time be his own; his poetry could not be hurried but had to emerge in its own way. So, while he loved teaching (and was much loved by his students), he limited his teaching at Berkeley to one semester a year. This basically provided his only income, apart from occasional reviews or commissioned writing. "My income," Thom wrote, "averages about half that of a local bus-driver or street sweeper, but it is of my own choosing, since I prefer leisure to working at a full-time job." But I do not think Thom felt too constrained by his slender means; he had no extravagances (though he was generous with others) and seemed naturally frugal. (Things eased up in 1992, when he received a MacArthur Award, and after this he was able to travel more and enjoy some financial ease, to indulge himself a bit.)

We often wrote to each other about books we were excited by or thought the other would like. ("The best new poet I've discovered in years is Rod Taylor . . . a far-out writer—have you read him yet?" I had not, but immediately got *Florida East Coast Champion*.) Our tastes did not always coincide, and one book which I had enthused about aroused his contempt and anger and criticism so fierce I was glad it was contained in a private letter. (Like Auden, Thom rarely reviewed what he did not like, and in general his reviews were written in the mode of appreciation.[2] I loved the generosity, the balance, of these critical writings, especially in *The Occasions of Poetry*.)

Thom was far more articulate than I when it came to commenting on each other's work. I admired almost all of his

---

2. Early in 1970, when Thom was due to be in New York, I told him that Auden was having a birthday party, as always, on February 21, and asked him if he would like to come along. He declined, and it was only in 1973, after Auden's death, that he said anything on the subject (in a letter of October 2, 1973): "Probably he was, apart from Shakespeare, the poet who most deeply influenced me, who made it seem most possible for me to write my self. I don't believe he liked me very much, or so I'm told, but that doesn't matter any more than if I were to find that Keats didn't like me."

poems but rarely attempted to analyze them, whereas Thom was always at pains to define, as he saw them, the particular strengths and weaknesses of whatever I sent him. Especially in our early days, I sometimes felt terrified of his directness— terrified, in particular, that he would find my writings, such as they were, muzzy, dishonest, talentless, or worse. I had feared his criticisms at the beginning, but from 1971 on, when I sent him *Migraine*, I was eager for his reactions, depended on them, and gave them more weight than those of anyone else.

In the 1980s, I sent Thom manuscripts of several essays that I wrote to complete *The Man Who Mistook His Wife for a Hat*. Some of these he liked very much (particularly "The Autist Artist" and "The Twins"), but one, "Christmas," he called "a disaster." (Ultimately, I agreed with him and consigned it to the dustbin.)

But the response which affected me most, for it contrasted what I had become with what I had been when I first met Thom, was contained in a letter he wrote to me after I sent him *Awakenings* in 1973. He wrote,

> *Awakenings* is, anyway, extraordinary. I remember when, some time in the late Sixties, you described the kind of book you wanted to write, simultaneously a good scientific book and worth reading as a well-written book, and you have certainly done it here. . . . I have also been thinking of the Great Diary you used to show me. I found you so talented, but so defi- cient in one quality—just the most important quality—call it humanity, or sympathy, or something like that. And, frankly, I despaired of your ever becoming a good writer, because I didn't see how one could be taught such a quality. . . . Your deficiency of sympathy made for a limitation of your observation. . . .

What I didn't know was that the growth of sympathies is something frequently delayed till one's thirties. What was deficient
in these writings is now the supreme organizer of *Awakenings*,
and wonderfully so. It is literally the organizer of your style,
too, and is what enables it to be so inclusive, so receptive, and so
varied. . . . I wonder if you know what happened. Simply working with the patients over so long, or the opening-up helped by
acid, or really falling in love with someone (as opposed to being
infatuated). Or all three . . .

I was thrilled by this letter, a bit obsessed, too. I did not
know how to answer Thom's question. I had fallen in love—
and out of love—and, in a sense, was in love with my patients
(the sort of love, or sympathy, which makes one clear-eyed). I
did not think that acid, of which I had had a fair sampling, had
played a real part in opening me up, though I knew that it had
been crucial for Thom.[3] (I was intrigued, though, to see that
the L-dopa I gave to the postencephalitic patients sometimes
produced effects similar to what I myself had experienced on
LSD and other drugs.) On the other hand, I felt that psychoanalysis had played a crucial role in allowing me to develop (I
had been in intensive analysis since 1966).

When Thom spoke of the growth of sympathy in one's thirties, I could not help wondering whether he was also thinking
of himself, in particular of the change in himself and his poetry
which one sees in *My Sad Captains* (he was thirty-two when

---

3. Thom wrote of this at length in his autobiographical essay "My Life
up to Now": "It is no longer fashionable to praise LSD, but I have no doubt
at all that it has been of the utmost importance to me, both as a man and
as a poet. . . . The acid trip is unstructured, it opens you up to countless
possibilities, you hanker after the infinite."

this was published), about which he later wrote, "The collection is divided into two parts. The first is the culmination of my old style—metrical and rational, but maybe starting to get a little more humane. The second half consists of taking up that human impulse . . . in a new form [which] almost necessarily invited new subject matter."

I was twenty-five when I first read *The Sense of Movement*, and what appealed to me then, along with the beauty of image and the perfection of form, was the almost Nietzschean emphasis on will. By the time I came to write *Awakenings*, in my late thirties, I had changed profoundly, and Thom had too. It was now his new poems, with their huge range of subject matter and sensibility, which appealed to me more, and we were both happy to leave the Nietzschean stuff behind. By the 1980s, as we both moved into our fifties, Thom's poetry, while never losing its formal perfection, grew freer and more tender. The loss of friends, surely, played a part here; when Thom sent me "Lament," I thought it the most powerful, the most poignant, poem he had ever written.

I loved the sense of history, of predecessors, in many of Thom's poems. Sometimes this was explicit, as in his "Poem After Chaucer" (which he sent me as a New Year's card in 1971); more often it was implicit. It made me feel at times that Thom *was* a Chaucer, a Donne, a Lord Herbert, who now found himself in the America, the San Francisco, of the late twentieth century. This sense of ancestors, of predecessors, was an essential part of his work, and he often alluded to, or borrowed from, other poets and other sources. There was no tiresome insistence on "originality," and yet, of course, everything he used was transmuted in the process. Thom later reflected on this in an autobiographical essay:

I must count my writing as an essential part of the way in which I deal with life. I am however a rather derivative poet. I learn what I can from whom I can. I borrow heavily from my reading, because I take my reading seriously. It is part of my total experience and I base most of my poetry on my experience. I do not apologize for being derivative. . . . It has not been of primary interest to develop a unique poetic personality, and I rejoice in Eliot's lovely remark that art is the escape from personality.

There is a danger, when old friends meet, that they will talk mostly of the past. Thom and I had both grown up in northwest London, been evacuated in the Second World War, played on Hampstead Heath, drunk in Jack Straw's Castle; we were both products of our families, schools, times, and cultures. This formed a certain bond between us and allowed an occasional sharing of recollections. But much more important was the fact that we had both been drawn to a new land, to the California of the 1960s, disenthralled from the past. We had launched on journeys, evolutions, developments, that could not be entirely predicted or controlled; we were constantly in motion. In "On the Move," which Thom wrote in his twenties, are the lines

> At worst, one is in motion; and at best,
> Reaching no absolute, in which to rest,
> One is always nearer by not keeping still.

Thom was still on the move, still full of energy, in his seventies. When I last saw him, in November of 2003, he seemed more intense, not less intense, than the young man of forty years earlier. Back in the 1970s, he had written to me, "I have just had *Jack Straw's Castle* published. I cannot guess what my

next book will be like." *Boss Cupid* was published in 2000, and
now, Thom said, he was getting ready for another book but had
no idea yet what it would be. He had, so far as I could judge, no
thoughts of slowing down or stopping. I think he was moving
forward, on the move, till the very minute he died.

I fell in love with Manitoulin, a large island in Lake Huron,
when I went there in the summer of 1979. I was still trying to work on my exasperating *Leg* book and had decided to
take off on an extended vacation where I could swim, think,
write, and listen to music. (I had only two tape cassettes, one
of Mozart's Mass in C Minor and the other of his Requiem. I
tend to get fixated on one or two pieces of music sometimes
and will play them again and again and again, and these were
the two pieces which had played in my mind five years earlier,
as I was slowly coming down the mountain with a useless leg.)

I wandered a lot around Gore Bay, the chief town on Manitoulin. I am normally rather shy, but I found myself opening conversations with strangers. I even went to the church
on Sunday because I enjoyed the feeling of community. As I
was preparing to leave, after an idyllic but not terribly productive six weeks, some of the elders in Gore Bay approached me
with an astonishing proposition. They said, "You seem to have
enjoyed your stay here; you seem to love the island. Our doctor has just retired after forty years. Would you be interested in
taking his place?" When I hesitated, they said that the province
of Ontario would give me a house and that—as I had seen—it
was a good life on the island.

I was greatly moved by this and thought about it for several
days, allowing myself to fantasize about being an island doctor.

But then, with some regret, I thought, this cannot work. I am not cut out to be a general practitioner; I need the city, clamorous though it is, and its large, diverse population of neurological patients. I had to say to the elders of Manitoulin, "Thank you—but no."

This was more than thirty years ago, but I still wonder, sometimes, how life would have been had I said yes to the elders of Manitoulin.

•

Later in 1979, I found a home on a very different island. I heard about City Island, a part of New York City, as soon as I started working at Einstein in the fall of 1965. Only a mile and a half long by half a mile wide, it had the feeling of a New England fishing village, and it felt a world apart from the Bronx, even though it was only ten minutes from Einstein and several of my colleagues lived there. The island had pleasant views of the sea in all directions, and going to one of its many fish restaurants for lunch provided a pleasant break in the day—a day which, if the research was challenging, might be eighteen hours long.

City Island had its own identity, rules, and traditions, and the natives of the island, the "clam diggers," seemed particularly respectful of idiosyncrasy, whether it was Dr. Schaumburg, a fellow neurologist who had had polio as a child, riding his big tricycle slowly up and down City Island Avenue, or Mad Mary, a woman who became psychotic at intervals and would stand in the back of her pickup truck, preaching hellfire. But Mary was accepted as just another neighbor. Indeed, she seemed to have a special role as a wise woman, a woman whose robust common sense and humor had been forged in the fires of psychosis.

When I was evicted from my apartment at Beth Abraham, I had rented the top floor of a house from a nice couple in Mount

Vernon, but I often drove or biked to City Island and Orchard Beach. I would bike down to the beach for a swim on summer mornings before work, and on weekends I would go for long swims, sometimes swimming around City Island, which took about six hours.

It was on such a swim, in 1979, that I spotted a charming-looking gazebo near the end of the island; I got out of the water to look at it and then strolled up the street, where I saw a "For Sale" sign in front of a little house. I knocked on the door, dripping, and met its owner—an ophthalmologist at Einstein. He had just completed his fellowship and was now moving, with his family, to the Pacific Northwest. He showed me around the house (I borrowed a towel so I would not drip inside), and I was hooked. Still in my swim trunks, I strode up City Island Avenue in my bare feet to the realtor's office and told her I wanted to buy the house.

I had yearned for a house of my own, such as I had rented in Topanga Canyon back in my UCLA days. And I wanted a house by the water so I could put on swim trunks and sandals and walk straight down to the sea. So the little red clapboard house on Horton Street, half a block from the beach, was ideal.

I had no experience of owning a house, and disaster quickly ensued. When I left the house that first winter for a week in London, I did not realize I had to leave the heat on to prevent the pipes from freezing. When I got back from London and opened the front door, I was greeted by an appalling sight. A pipe upstairs had burst and flooded, and the entire ceiling of my dining room hung in tatters over the dining room table. The table and chairs were totally ruined, as was the carpet beneath them.

While I was in London, my father had suggested that I

take his piano, now that I had a house; it was a beautiful old Bechstein grand dating from 1895, the year of his birth. He had had it for more than fifty years and had played it daily, but now, in his mid-eighties, his hands were getting crippled by arthritis. A wave of horror passed over me when I saw the devastation, made sharper by the thought that this is where the piano would have been had I got the house earlier in the year.

Many of my neighbors on City Island were sailors. The house next door belonged to Skip Lane and his wife, Doris. Skip had captained large merchant ships for most of his life, and he had a house so full of ship's compasses and steering wheels, binnacles and lanterns, that it had come to resemble a ship itself. The walls were covered by photographs of vessels he had commanded.

Skip had innumerable yarns of life at sea, but now that he was retired, he had given up his huge ships for a tiny, one-man Sunfish; he often sailed it across Eastchester Bay and thought nothing of sailing it all the way to Manhattan.

Though he must have weighed close to 250 pounds, Skip was enormously strong and amazingly agile. I would often see him fixing things on the roof of his house—I think he liked the feeling of being aloft—and on one occasion, when challenged, he climbed up a thirty-foot pylon of the City Island Bridge, hauling himself up by sheer muscle and then balancing on one of its girders.

Skip and Doris were perfect neighbors, never intrusive, but endlessly helpful where needed, and with a great energy and gusto for life. There were only a dozen or so houses on Horton Street, perhaps thirty of us in all, and insofar as we had a leader, a man of decision, it was Skip.

At one point in the early 1990s, we were warned that a big

hurricane was coming our way, and police came with bullhorns telling us to evacuate. But Skip, who knew all the vagaries of storms and sea, and had a voice louder than any police bullhorn, disagreed. "Avast!" he roared. "Stay put!" He invited us all to a hurricane party on his porch at noon, to watch the eye of the hurricane pass over. Just before noon, as Skip had predicted, the wind died down, and a sudden calm and quietness descended. Now, in the eye of the hurricane, the sun shone, the sky was clear—a halcyon, magical calm. Skip told us one could some-times see birds or butterflies which had been carried thousands of miles, even from Africa, in the eye of the storm.

No one on Horton Street locked their doors. We all looked out for one another and for the little beach we shared. It may have been only a few yards wide, but it was *our* beach, and every Labor Day we had a party on the tiny patch of sand, with a whole pig roasting slowly on a spit.

I often went for long swims in the bay with another neigh-bor, David, who had the caution and common sense which I lacked and, on the whole, kept me out of trouble. But I some-times went too far; once I swam all the way out to the Throgs Neck Bridge and was almost cut in half by a boat. David was shocked when I told him about this and said that if I insisted on swimming ("like an idiot") in shipping lanes, I should at least tow a bright orange float behind me for visibility.

I sometimes encountered little jellyfish in the waters off City Island. I ignored the slight burning as they brushed against me, but in the mid-1990s much larger jellyfish started to appear: *Cyanea capillata,* the lion's mane jellyfish (like the one responsible for a mysterious death in the final Sher-lock Holmes story). It was not good to brush against these. They would leave agonizing welts across the skin, along with frightening effects on heart rate and blood pressure. On one occasion, the ten-year-old son of one of my neighbors had a dangerous anaphylactic reaction to being stung; his face and tongue became so swollen that he could hardly breathe, and only a prompt injection of adrenaline saved him.

When the jellyfish plague got worse, I took to swimming in full scuba gear, including a face mask. Only my lips were exposed, and I slathered them with Vaseline. Even so, I got the horrors one day when I found a *Cyanea* the size of a football in one of my armpits; this was the end of carefree swimming for me.

Every May and June, during the full moon, an ancient and marvelous ceremony was enacted on our beach, as on beaches all over the Northeast, when the horseshoe crabs, creatures little changed from the Paleozoic, clambered slowly up to the water's edge for their annual mating. Watching this ritual, which had occurred every year for more than 400 million years, I got a vivid feeling for the reality of deep time.

•

City Island was a place to saunter, to stroll around slowly—up and down City Island Avenue, and into its cross streets, each only one or two blocks long. There were many fine old gabled houses going back to Victorian days, and there were still a few shipyards left from its heyday as a yacht-building center. City Island Avenue was lined with seafood restaurants, ranging from the long-established, elegant Thwaite's Inn to Johnny's Reef Restaurant, an open-air fish and chips place. My own favorite, quiet and unpretentious, was Spouter's Inn, with whaling pictures on the walls and pea soup every Thursday. It was Mad Mary's favorite place, too.

Much of my shyness melted away in this small-town atmosphere. I was on easy first-name terms with the manager of Spouter's, with the man who ran the gas station, and with the clerks at the post office (they said no one in living memory had sent or received so many letters, and this increased by an order of magnitude when *Hat* was published).

Sometimes, oppressed by the emptiness and silence of the house, I would go to the Neptune, a curiously uncrowded and unpopular restaurant at the end of Horton Street, and sit there for hours, writing. I think they quite liked their quiet writer, who would order a different dish every half hour or so, because he did not want the restaurant to lose money on his account.

•

Early in the summer of 1994, I was adopted by a stray cat. I got back from the city one evening, and there she was, sitting sedately on my porch. I went into the house and brought out a saucer of milk; she lapped thirstily. Then she looked at me, a look that said, "Thanks, buddy, but I'm hungry, too."

I refilled the saucer and came back with a piece of fish, and this sealed an unspoken but clear covenant: she would stay with me, if we could arrange a way of living together. I found a basket for her and put it on a table on my front porch, and the next morning, I was happy to see, she was still there. I gave her more fish, left a bowl of milk for her, and took off for work. I waved good-bye to her; I think she understood that I would be back.

That evening, she was there awaiting me; indeed, she greeted me by purring, arching her back, and rubbing herself against my leg. I felt oddly touched when she did this. After the cat had eaten, I settled myself, as I liked to, on a sofa by the porch window, to eat my own dinner. The cat jumped up on her table outside and watched me as I ate.

When I got back the next evening, I put her fish out on the

floor again, but this time, for some reason, she would not eat it. When I put the fish on the table, she jumped up, but it was only when I had settled myself on the sofa by the window that the cat, lying parallel to me, now started to eat her supper as I ate mine. So we ate together, in synchrony. I found this ritual, which was to be repeated every evening, remarkable. I think we both had a feeling of companionship—which one might expect with a dog but rarely with a cat. The cat liked to be with me; she would even, after a few days, walk down to the beach with me and sit next to me on a bench there.

I do not know what she did in the day, though once she brought me a small bird and I realized she must have been hunting, as cats do. But whenever I was in the house, she would be on the porch. I was charmed and fascinated by this interspecies relationship. Was this how man and dog had met a hundred thousand years ago?

When it got cooler, in late September, I gave the cat—I just called her Puss, and she responded to this—to friends, and Puss lived happily with them for the next seven years.

•

I was lucky enough to find Helen Jones, a wonderful cook and housekeeper who lived close by, and she would come to me once a week. When she arrived each Thursday morning, we would set out for the Bronx to do some shopping together, our first stop being a fish shop on Lydig Avenue run by two Sicilian brothers who were as like as twins.

When I was a boy, the fishmonger would come to our house every Friday carrying a bucket swimming with carp and other fish. My mother would boil the fish, season and grind them all together, and make a great bowl of gefilte fish; this, along with salads, fruits, and challahs, kept us going through the Sabbath, when cooking was not allowed. The Sicilian fishmongers on Lydig Avenue were happy to give us carp, whitefish, and pike. I had no idea how Helen, a good churchgoing Christian, would manage with making such a Jewish delicacy, but her powers of improvisation were formidable, and she made magnificent gefilte fish (she called it "filter fish") which, I had to acknowledge, was as good as my mother's. Helen refined her filter fish each time she made it, and my friends and neighbors got a taste for it, too. So did Helen's church friends; I loved to think of her fellow Baptists gorging on gefilte fish at their church socials.

•

One summer day in the 1990s, when I returned from work, I encountered a strange apparition on my porch, a man with an enormous black beard and head of hair—a mad tramp, was my immediate impression. Only when the tramp spoke did I realize who he was—my old friend Larry. I had not seen him for many years and had come to think, as many of us had, that he was probably dead.

I had met Larry early in 1966, when I was trying to recoup from my first, evil drug-addicted months in New York. I was eating well, exercising and gaining my strength back, going regularly to a gym in the West Village. The gym opened at 8:00 on Saturday mornings, and I was often the first person there. One Saturday, I started my workout with the leg-press machine; I had been a mighty squatter when I was in California and wondered how much of my strength had come back. I worked the weight up to 800 pounds—easy; 1,000 pounds— challenging; 1,200 pounds—folly. I knew it was too heavy for me, but I refused to concede failure. I did three reps, four, just, and my strength gave out on the fifth. I lay helpless with 1,200 pounds on top of me, my knees crushed into my chest. I could hardly breathe, much less shout for help, and started to wonder how long I could hold out. I felt my head getting engorged with blood, and I feared a stroke was imminent. At that moment, the door swung open, and a powerful young man walked in, saw my predicament, and helped me up with the bar. I hugged him and said, "You saved my life."

Despite his quick action, Larry seemed very shy. It was difficult for him to make contact, and he had a driven, anxious look, his eyes never still. But now that contact had been made, he could hardly stop talking; perhaps I was the first soul he had spoken with in weeks. He was nineteen, he told me, and the previous year he had been discharged from the army as mentally unstable. He lived on a small pension from the government. He subsisted, as far as I could tell, on milk and bread; he spent sixteen hours a day walking the streets (or running, if he was in the country) and was happy to bed down anywhere at night.

He had never known his parents, he told me. His mother had advanced multiple sclerosis by the time he was born and was physically incapable of looking after him. His father was an alcoholic who abandoned them soon after Larry was born, and Larry had been farmed out to a series of foster parents. It seemed to me that he had never known any real stability.

I did not care to make a "diagnosis" of Larry, even though I was very free with psychiatric terms in those days. All I could think of was how much love and care and stability had been denied to him, how much respect had been denied to him, and I marveled that he had survived psychically at all. He was very intelligent and far better informed about current affairs than I was. He would find old newspapers and read them from cover to cover. He thought tenaciously, relentlessly, about everything he had read or been told. He would take nothing on trust.

He had no intention of ever getting a job, and this, I thought, took a special sort of integrity. He was determined to avoid a meaningless busyness; he was frugal, and he could live and even save on his modest pension.

Larry spent his days walking, and it was not unusual for him to walk the twenty miles from his apartment in the East Village to my house in City Island. He sometimes stayed over on my living room couch, and one day at the bottom of the refrigerator I found some very heavy bars, bars of gold that Larry had bought over the years. He had stashed them in my house, feeling they would be safer there than in his apartment. Gold, he said, was the only possession one could trust in an unstable world; shares, bonds, land, art, could all lose their value overnight, but gold ("element 79," he used to say, to please me) would always hold its value. Why should he work, hold a job, when he could live, be independent, a free man, without doing so? I liked his courage, his forthrightness, in saying this and felt, in a way, that he was one of the freest souls I knew.

Larry was transparent and sweet natured, and many women found him attractive. He had been married for some years to a generously proportioned woman in the East Village, but, hideously, she was murdered one day by thugs who broke into their apartment to find drugs. They found none, but Larry found her corpse.

Larry had always lived largely on milk and bread, and in his distress now over her death he wanted nothing but milk. He became consumed by a fantasy of traveling the world with an enormous, lactating woman who would cradle him like a baby and let him suckle at her breast. I never heard a more primal fantasy.

Sometimes I would not see Larry for weeks or months—I had no way of contacting him—but then he would suddenly reappear.

He was an alcoholic like his father, and alcohol set off something mischievous and self-destructive in his brain. He knew this and usually avoided drinking. In the late 1960s, we dropped acid together a couple of times, and he liked coming with me, riding on the back of my motorbike, to visit my cousin Cathy— one of Al Capp's daughters—who lived in Bucks County. Cathy was schizophrenic, but she and Larry understood each other intuitively and formed a strange bond.

Helen, too, adored Larry, and all of my friends liked him; he was an utterly independent human being, a sort of modern, urban Thoreau.

In New York, I got to know some of my American cousins, the Capps (their original name was Caplin, and they were second cousins, really). The eldest was Al Capp, the cartoonist. He had two younger brothers—Bence, also a cartoonist, and Elliott, a cartoonist and playwright—and a sister, Madeline.

I have vivid memories of the first Capp family seder I attended in 1966. I was thirty-two, and Louis Gardner, Madeline's husband, was a young and handsome forty-eight, very upright, with a military bearing; he was a colonel in the reserves, as well as an architect. Louis, at the head of the table, conducted the seder, with Madeline at the other end and an extraordinary pack of family members in between—Bence, Elliott, and Al, with their wives. Louis and Madeline's children were running all over the place, when they were not reciting the four questions or looking for the *afikomen*.

We were all in our prime then. Al, still the brilliant and beloved creator of *Li'l Abner*, was read and admired all over America. Elliott, the most thoughtful of the brothers, was admired for his essays and plays. Bence (Jerome) crackled with creative energy, and Madeline, the darling of her brothers, was the center of it all. They were all brilliant, exuberant talkers, and I sometimes thought Madeline was the smartest of them all; the stroke which was to leave her aphasic was still years in the future.[4]

I saw quite a lot of Al, who was a strange figure when I met him in the mid-1960s. All of the brothers had been communists or fellow travelers in the 1930s, but Al went through a strange political reversal in the 1960s, when he became a friend of Nixon's and Agnew's (though not entirely trusted by them, I suspect, because his wit and satire might be aimed at anyone in power).

Al had lost his leg in a traffic accident as a boy of nine, and he sported a very massive wooden leg (it made me think of Captain Ahab's whalebone ivory leg). It may be that some of his aggressiveness, some of his competitiveness, some of his blatant sexuality, had to do with having been maimed, feeling that he had to show he was not a cripple but a sort of superman, but I never encountered this aspect of Al. He was always

---

4. Madeline had this stroke when she was only about fifty. She was aphasic ever after this, but she was aphasic with such wit, such style, and such ingenuity that she gave aphasia a new meaning.

friendly and genial with me, and I grew fond of him and thought him full of creative vitality and charm.

In the early 1970s, Al, besides his cartooning, did a lot of university lecturing. He was a brilliant speaker and a darling of the lecture circuit, although dark rumors started to collect around him—that he was, perhaps, a little too forward with some of the women students. The rumors got darker; accusations were leveled. There was a scandal, and Al was fired by the hundreds of syndicated papers that he had worked for all his life. Suddenly the beloved cartoonist who had created Dogpatch and the Shmoo, who was in some ways the graphic Dickens of America, found himself reviled and out of a job.

He retreated for a while to London, where he lived in a hotel and published occasional articles and cartoons. But he was a broken man, as they say; his rambunctiousness, his vitality, deserted him. He remained depressed, and in declining health, until his death in 1979.

·

Another cousin, Aubrey "Abba" Eban, was the prodigy of the family, the brilliant elder son of my father's sister, Alida. He had shown exceptional gifts as a boy and had gone on to a dazzling career at Cambridge, becoming president of the Cambridge Union, gaining a triple first, and going on to become a lecturer there in Oriental languages. He had shown that despite the anti-Semitism prevalent in the England of the 1930s, a Jewish boy with no advantages of wealth or birth or social connections, with nothing except an extraordinary brain, could make it to the top in one of England's oldest universities.

His passionate eloquence and great wit were already fully developed by the time he was twenty, but it was still unclear whether this would lead him to a political life—his mother, my aunt, had translated the Balfour Declaration into French and Russian in 1917, and Aubrey had been a committed and idealistic Zionist since childhood—or whether he would remain a scholar at Cambridge. The war and the developments in Palestine determined his course.

Aubrey was nearly twenty years my senior, and I did not have much contact with him until the mid-1970s. His life was in Israel; mine was in England and then the United States; his was the life of a diplomat and politician, mine the life of a physician and scientist. We saw each other rarely and briefly at family weddings and other events. And when Aubrey did visit New York, as foreign minister or deputy prime minister of Israel, he always seemed to be surrounded by security men, and there was little chance of saying more than a few words.

But one day in 1976, we were both invited to lunch by Madeline, and as soon as Aubrey and I met, it was evident to both of us, and to everyone there, that we showed a startling similarity of gesture and posture—the way we sat, our abrupt bulky movements, our style of speech and mind. At one point, we both got up suddenly from opposite ends of the table and collided over the beetroot jelly, which we both loved but everyone else hated. The whole table was set laughing by these similarities and coincidences, and I said to Aubrey, "I have hardly ever met you, and our lives are quite different, but I have the feeling that there is more genetic similarity between the two of us than there is between me and my three brothers." He said that he had had the same feeling, that I was in some way closer to him than his three siblings.

How can this be? I asked. "Atavism" was his instant reply.

"Atavism?" I blinked.

"Yes, *atavus*, a grandfather," Aubrey replied. "You never

knew our grandfather Elivelva (even though you have the
same Hebrew and Yiddish name). He died before you were
born. But I was brought up by him when we came to England.
He was my first real teacher. People laughed when they saw us
together; they said there was an uncanny similarity between
the old man and the child. There was no one else in his generation who spoke or moved or thought the way he did, no one
at all like him in the parental generation, and I thought there
was no one like him in my own generation, until you came
through the door, and I thought my grandfather had come to
life."

There was an element of tragedy, or paradox, in store for
Aubrey, who had won the world's ear as "the voice of Israel."
His passionate and polished eloquence, his Cambridge accent,
came to be seen by a new generation as pompous and old-
fashioned, and his fluency in Arabic and sympathetic knowledge of Arab culture (his first book had been a translation of
Tawfiq al-Hakim's *Maze of Justice*) rendered him almost suspect in an increasingly partisan atmosphere. So eventually he
fell from power and returned to life as a scholar and historian
(as well as becoming a brilliant expositor in books and on television). His own feelings, he told me, were mixed: he felt "a
void" after decades of intense immersion in politics and diplo-
macy but he also felt a sudden, unprecedented peace of mind.
His first act, as a free man, was to go for a swim.

Once, while Aubrey was a visiting professor at the Institute for Advanced Study in Princeton, I asked him how academic life suited him. He looked wistful and said, "I pine for
the arena." But as the arena became stormier and narrower
and more partisan, Aubrey, with his wide cultural sympathies
and his spaciousness of mind, pined for it less and less. I once

asked him how he wanted to be remembered, and he said, "As a teacher."

Aubrey loved telling stories, and knowing of my interest in the physical sciences, he told me several stories of his contacts with Albert Einstein. Following Chaim Weizmann's death in 1952, Aubrey had been delegated to invite Einstein to become the next president of Israel (Einstein, of course, declined). On another occasion, Aubrey recounted with a smile, he and a colleague from the Israeli consulate visited Einstein in his house in Princeton. Einstein invited them in and courteously asked if they would like coffee, and (thinking that an assistant or housekeeper would make it), Aubrey said yes. But he was "horrified," as he put it, when Einstein trotted into the kitchen himself. They soon heard the clatter of cups and pots and an occasional piece of crockery falling, as the great man, in his friendly but slightly clumsy way, made the coffee for them. This, more than anything, Aubrey said, showed him the human and endearing side of the world's greatest genius.

During the 1990s, no longer burdened or exalted by office, Aubrey would come to New York in a much freer and easier way, and I saw him more frequently, sometimes with his wife, Suzy, and often with his younger sister, Carmel, who also lived in New York. Aubrey and I became friends, the great difference in our lives and the near twenty-year difference in our ages mattering less and less.

Dear, monstrous Carmel! She outraged everyone, at least all her family, but I had a soft spot for her.

For many years, Carmel was a mythical figure, an actress somewhere in Kenya, but in the 1950s she came to New York, married a director called David Ross, and with him established a small theater to stage his favorite Ibsen and Chekhov plays (though her own preference was always for Shakespeare).

When I met her, in May of 1961, I had just ridden from San Francisco on my motorbike—the secondhand bike that blew up in Alabama—and made the rest of the journey to New York by hitchhiking. I was rather dirty and disheveled when she received me in their elegant Fifth Avenue apartment; she ordered me to have a bath, and clean clothes were procured while mine were laundered.

David was riding high at the time; he had had a series of critical and popular successes and, Carmel told me, was beginning to be seen as a major figure in the New York theater world. He was in a flamboyant, extravagant mood when I saw him; he bellowed, he roared like a lion, and he took us to an unbelievably expensive, six-course dinner at the Russian Tea Room—everything on the menu, with half a dozen assorted vodkas thrown in. This went beyond mere exuberance, and I wondered if there was a touch of mania in him.

Carmel, too, was pretty high; she saw no reason why she could not master Norwegian and Russian—with her ear for languages, it should take only a few weeks—and provide her own translations of Ibsen and Chekhov. Her translation might have been part of the reason David's *John Gabriel Borkman* crashed and lost a good deal of money when it opened in London. Carmel had wheedled most of this money from her family, who could ill afford it, and she never paid it back. Some years later, David had to be hospitalized in New York—he was prone to severe depressions—and soon after he died, whether from an accidental overdose or suicide was never clear. Carmel, deeply shaken, returned to London, where she had family and friends.

•

Carmel and I would meet again in 1969, while I was in London writing the first case histories of *Awakenings* and *Migraine* was still in press with Faber & Faber. Carmel asked to see what I had written, and after reading the proofs of *Migraine,* she said, "Why, you are a writer!" No one had ever quite said this to me before; *Migraine* was being published by the medical division of Faber & Faber and was seen by them as a medical book, an idiosyncratic monograph on migraine—not as "writing." And no one yet had seen the first case histories of *Awakenings,* no one but Faber & Faber, who rejected them as unpublishable. So I was buoyed by Carmel's words and by her thought that *Migraine* might be received well not only by the medical profession but by general and even "literary" readers as well.

When Faber & Faber delayed publication of *Migraine,* I grew more and more frustrated, and Carmel, seeing this, stepped in decisively.

"You must get an agent," she said. "Someone who will stand up for you, who won't let you be screwed."

It was Carmel who introduced me to Innes Rose, the agent who put pressure on my publishers to release the book. Without Innes, without Carmel, *Migraine* might not have seen the light of day.

Carmel returned to New York in the mid-1970s, after her mother's death, and got an apartment on East Sixty-Third Street. She acted as a sort of agent for me and for Aubrey, who was involved then in a series of books and television programs on the history of the Jews. But neither agenting nor acting, both part-time, could pay Carmel's rent in an ever more expensive New York, so Aubrey and I made up the shortfall together and continued to do so for the next thirty years.

Carmel and I saw a lot of each other in those years. We often went to plays together, and one play we saw was *Wings*, in which Constance Cummings played an aviatrix who loses language after a stroke. Carmel turned to me at one point and asked if I did not find her performance deeply moving and was taken aback when I said no.

Why not? she demanded. Because, I replied, her speech sounded nothing like the speech of people with aphasia.

"Oh, you neurologists!" Carmel said. "Can't you forget your neurology for a while and let yourself be swept along by the drama, the acting?"

"No," I said. "If the language sounds nothing like aphasia, then the whole play seems unreal to me." She shook her head at my narrow-mindedness and intransigence.[5]

Carmel was excited when *Awakenings* was taken up by Hollywood and I met Penny Marshall and Robert De Niro. But her instinct played her wrong on my fifty-fifth birthday, when De Niro attended my party on City Island and (in the invisible way he has) managed to come to my little house and ensconce himself quietly upstairs without anyone recognizing him. When I told Carmel that De Niro had arrived, she said, very loudly, "That's not De Niro. He's a look-alike, a double, sent by the studio. I know what a real actor is like, and he doesn't take me in for a minute." She knew how to project her voice, and everyone heard her comment. I myself became uncertain and went down to the phone box on the corner, where I phoned De Niro's office. Puzzled, they said of course that was the real De

---

5. We went backstage to see Cummings after the play, and I asked her whether she had met many people with aphasia. "No, not a single one," she replied. I said nothing, but I thought, "It shows."

Niro. And no one was more amused than De Niro himself, who had heard Carmel's bellowing.

·

Dear, monstrous Carmel! I enjoyed her company—when she was not infuriating me. She was brilliant, funny, a wickedly talented imitator; she was impulsive, ingenuous, feckless, but she was also a fantast, a hysteric, and a leech, always sucking more and more money from everyone around her. She was a dangerous houseguest (I learned later), swiping art books from her hosts' libraries and selling them to secondhand bookshops. I often thought of our Auntie Lina, who blackmailed rich people to give money to the Hebrew University. Carmel never blackmailed anyone, but she resembled Lina in a number of ways: she too was a monster, hated by some of the family, but someone I had a soft spot for. Carmel was not unaware of the resemblance.

When Carmel's father died, he left the bulk of his estate to her, because he recognized that she was the neediest of the children. Any resentment her brothers and sister felt was partly balanced by the feeling that now, with her patrimony, she was set for life, provided she lived sensibly and avoided follies or extravagance; she would no longer have to scrounge or be supported by them. I too was pleased that I would no longer feel obliged to send her a monthly check.

But she had other ideas; she had missed being a part of the theater world since David's death. Now she had money and could produce, direct, and act in a favorite play herself; she chose *The Importance of Being Earnest*, which would allow her to star as Miss Prism. She rented a theater, assembled a cast, and organized publicity, and, as she hoped, the performance was a success. But then, in the mysterious way things happen, there was no follow-up. She had blown every cent of her inheritance in a single mad, idiotic gesture. The family was enraged, and she was, once again, broke.

Carmel took it all rather cheerfully, even though it was, in a sense, a repetition of what happened with *John Gabriel Borkman* thirty years before. But she was less resilient now. She was seventy, though she looked younger; she had diabetes, which she was careless with; and the family (except for Aubrey, who always stood by her, however much she infuriated him) was no longer on speaking terms with her.

Aubrey and I resumed our monthly checks, but something, at a deeper level, was broken inside Carmel. She felt, I think, that she had had her last chance for Broadway glory and stardom. Her health deteriorated, and this forced her into an assisted-living residence. She sometimes became delusional, whether from incipient dementia or diabetes or both, and she was occasionally found, disheveled and disoriented, wandering the streets near the Hebrew Home. At one point, she was convinced she was starring with Tom Hanks in a movie directed by Steven Spielberg.

But there were other days without mishaps, when she enjoyed outings to the theater—her first and last love—and walks in the lovely Wave Hill gardens, close to the Hebrew Home. She decided at this point to write an autobiography; she wrote easily and well and had an unusual, exotic life story to tell. But her autobiographic memory was beginning to fail her, as the dementia stealthily advanced.

Her "performance" memory, her actor's memory, in contrast, was untouched. I had only to prompt her with the beginning of any Shakespeare speech and she would continue the speech, becoming Desdemona, Cordelia, Juliet, Ophelia, whoever—completely taken over by the character she was enacting. The nurses, who usually saw her as an ill, demented old woman, were stunned by these transformations. Carmel once said to me that she had no identity of her own, only those of the characters she played—this was an exaggeration, for she had plenty of character and ego in her earlier days—but now, with the leaching of her own identity by dementia, this was almost literally the case; she only came to, became a full person, in those minutes when she became Cordelia or Juliet.

The last time I visited, she had pneumonia; her breathing was fast, irregular, rasping. Her eyes were open but unseeing; she did not blink when I waved a hand near them, but I thought that perhaps she might still hear and recognize a voice.

I said, "Good-bye, Carmel," and a few minutes later she was dead. When I phoned her brother Raphael to tell him of her death, he said, "God rest her soul—if she had one."

Early in 1982, I received a packet from London containing a letter from Harold Pinter and the manuscript of a new play, *A Kind of Alaska*, which, he said, had been inspired by *Awakenings*. In his letter, Pinter wrote that he had read *Awakenings* when it originally came out in 1973 and had thought it "remarkable." He had wondered about its dramatic possibilities, but then, seeing no clear way to go forward, he had put the subject out of his mind, until it suddenly came back to him eight years later. He had awoken one morning the previous summer with the first words of the play clear and pressing in his mind: "Something is happening." The play had then rapidly "written itself," he said, in the days that followed.

A Kind of Alaska is the story of Deborah, a patient who has been in some deeply strange, inaccessible frozen state for twenty-nine years. She awakens one day and has no idea of her age or what has befallen her. She thinks the grey-haired woman near her is some sort of cousin or "an aunt I never met," and the revelation that this is her younger sister shocks her into the reality of her situation.

Pinter had never seen our patients or the documentary of *Awakenings,* though my patient Rose R. had clearly been the model for his Deborah. I imagined Rose reading the play and saying, "My God! He's got me." I felt Pinter had somehow perceived more than I had written; he had divined, inexplicably, a deeper truth.

In October of 1982, I went to the play's opening at the National Theatre in London. Judi Dench gave a remarkable performance as Deborah. I was amazed at this, as I had been amazed at the verisimilitude of Pinter's conception, for Dench, like Pinter, had never met a postencephalitic patient. Indeed, she said, Pinter had forbidden her to do so as she prepared for her role; he felt that she should create the character of Deborah entirely from his lines. Her performance was gripping. (Later, however, Dench did see the documentary and visit some of the postencephalitics at the Highlands Hospital, and I felt that her performance after this, while perhaps more realistic, was less gripping. Perhaps Pinter was right.)

Up to this point, I had had reservations about dramatic representations or anything else "based on," or "adapted from," or "inspired by" my own work. *Awakenings* was the real thing, I felt; anything else would surely be "unreal." How could it be real if it lacked direct firsthand experience with the patients? Yet Pinter's play showed me how a great artist can re-confer, reimagine, reality. I felt Pinter had given me as much as I gave him: I had given him a reality, and he had given me one back.[6]

•

In 1986, I was in London when I was approached by the composer Michael Nyman—how would I feel, he asked, about a "chamber opera" based on the title story of *The Man Who Mistook His Wife for a Hat*? I said I could not imagine such a thing, and he replied that I had no need to; *he* would imagine it. He had in fact already done so, for the next day he presented me with a score and spoke of a librettist he had in mind, Christopher Rawlence.

I spoke with Chris at length about Dr. P. and finally said that I could not agree to an opera without the approval of his widow. I suggested to Chris that he meet her and gently enquire how she might feel about a possible opera (both she and Dr. P. had been opera singers themselves).

Chris went on to form a very warm, cordial relationship with Mrs. P., and she plays a much larger role in the opera than she does in my account. Nonetheless, I was very tense when the opera was first performed in New York. Mrs. P. came to the opening, and I kept glancing at her, fearfully misinterpreting every expression on her face. But after the performance, she came up to the three of us—Michael, Chris, and me—and said, "You have done honor to my husband." I loved that; it made me feel we had not taken advantage of him or misrepresented his situation.

---

6. I have since felt this about other works inspired by my own—above all by Peter Brook's brilliant theatrical presentations of *L'Homme Qui . . .* in 1993 and *The Valley of Astonishment* in 2014, and by a ballet inspired by *Awakenings*, with music by Tobias Picker.

•

In 1979, two young film producers, Walter Parkes and Larry
Lasker, approached me. They had read *Awakenings* a few years
earlier in an anthropology class at Yale and hoped to turn it into
a feature film. They visited Beth Abraham and met many of the
postencephalitic patients, and I agreed to let them develop a
script. Several years then passed in which I heard nothing.

I had nearly forgotten about the project when they contacted me again eight years later, saying that Peter Weir had
read *Awakenings* and the script inspired by it; he was very
interested in directing it. They sent me the script, written by
a young writer called Steve Zaillian, and this arrived on Hal-
loween in 1987, the day before I was due to meet Peter Weir.
I hated the script, especially an invented subplot in which
the doctor figure falls in love with a patient, and I said so, in
no uncertain terms, to Weir when he arrived. He was taken
aback, understandably, though he understood my position. A
few months later, he withdrew from the project, saying that he
saw all sorts of "reefs and shoals" and did not feel he could do
it justice.

Over the next year, the script went through many refine-
ments, as Steve, Walter, and Larry worked to produce something which would remain true to the book and the patients'
experiences. Early in 1989, I was told that Penny Marshall
would direct the film and that she would be visiting me with
Robert De Niro, who would play the patient Leonard L.

I was not quite sure how I felt about the script, for while
in some ways it aimed at a very close reconstruction of how
things had been, it also introduced several subplots which were
entirely fictional. I had to renounce the notion that it was, in
any way, "my" film: it was not my script, it was not my film,

it would largely be out of my hands. It was not entirely easy to say this to myself, and yet it was also a relief. I would be able to advise and consult, to ensure medical and historical accuracy; I would do my best to give the film an authentic point of departure, but I would not have to feel responsible for it.[7]

•

Robert De Niro's passion to understand what he is going to portray, to research it in microscopic detail, is legendary. I had never before witnessed an actor's investigation of his subject—the investigation that would finally culminate in the actor's *becoming* his subject.

By 1989, nearly all the postencephalitic patients at Beth Abraham had died, but there were still nine remaining at the Highlands Hospital in London. Bob felt it important to visit them, so we went to see them together. He spent many hours talking with the patients and making research tapes which he could study at length. I was impressed and moved by his powers of observation and empathy, and I think the patients themselves were moved by a sort of attention they had rarely met before. "He really observes you, looks right into you," one of them said to me the following day. "Nobody's really done that since Dr. Purdon Martin. *He* tried to understand what was really going on with you."

On my return to New York, I met Robin Williams, who was

7. The documentary film of *Awakenings* was studied in minute detail by all the actors who would play postencephalitics; this became the primary visual source for the feature film, along with the miles of Super 8 film and audiotape which I had recorded myself in 1969 and 1970.

The documentary had never been broadcast outside the U.K., and the release of the Hollywood film seemed an ideal time to offer it to PBS. But Columbia Pictures insisted that we not do this; it thought it might distract from the "authenticity" of the feature film, an absurd idea.

to play the doctor—me. Robin wanted to see me in action, interacting with the sorts of patients I had worked and lived with in *Awakenings*, so we went off to the Little Sisters of the Poor, where there were two postencephalitic patients on L-dopa whom I had followed for several years.

A few days later, Robin came with me to Bronx State. We had spent a few minutes in a very disturbed geriatric ward, where half a dozen patients were shouting and talking bizarrely all at once. Later, as we drove away, Robin suddenly exploded with an incredible playback of the ward, imitating everyone's voice and style to perfection. He had absorbed all the different voices and conversations and held them in his mind with total recall, and now he was reproducing them, or, almost, being possessed by them. This instant power of apprehension and playback, a power for which "mimicry" is too feeble a word (for they were imitations full of sensitivity, humor, and creativity), was developed to an enormous degree in Robin. But it constituted, I came to think, only the first step in his actorial investigation.[8]

I was soon to find myself the subject of his investigation. After our first few meetings, Robin began to mirror some of my mannerisms, my postures, my gait, my speech—all sorts of things of which I had been hitherto unconscious. It was disconcerting to see myself in this living mirror, but I enjoyed being with Robin, driving around, eating out, laughing at his

8. This reminded me of how, a couple of years earlier, I had had a visit from Dustin Hoffman, who was researching his role as an autistic man in the film *Rain Man*. We visited a young autistic patient of mine at Bronx State and then went for a stroll in the botanical garden. I was chatting with Hoffman's director, and Hoffman was following a few yards behind. Suddenly I thought I heard my patient. I was extremely startled and turned round and saw it was Hoffman thinking to himself, but thinking with his voice and body, thinking enactively.

incandescent, rapid-fire humor, impressed by his wide range of knowledge.

A few weeks later, as we were chatting in the street, I got into what I am told is a characteristic pensive pose, and I suddenly realized that Robin was in exactly the same posture. He was not imitating me; he had become me, in a sense; it was like suddenly acquiring a younger twin. This disquieted both of us a bit, and we decided that there needed to be some space between us so that he could create a character of his own—based on me, perhaps, but with a life and personality of its own.[9]

·

I took the cast and crew to Beth Abraham several times to get the atmosphere and mood of the place, and most especially to see patients and staff who remembered the events of twenty years before. At one point, we invited all of the doctors, nurses, therapists, and social workers who had worked with the post-encephalitics in 1969 for a sort of reunion. Some of us had long since left the hospital, and some of us had not seen each other for years, but that evening in September we swapped memories of the patients for hours, each person's memories triggering others. We realized again how overwhelming, how historic

---

9. Over the next twenty-five years, Robin and I became good friends, and I grew to appreciate—no less than the brilliance of his wit and his sudden, explosive improvisations—his wide reading, the depth of his intelligence, and his humane concerns.

Once, when I gave a talk in San Francisco, a man from the audience asked me an odd question: "Are you English or are you Jewish?"

"Both," I replied.

"You can't be both," he said. "You have to be one or the other."

Robin, who was in the audience, brought this up at dinner afterwards and, using an ultra-English, Cambridge voice laced with Yiddish and Yiddish aphorisms, gave a stunning demonstration of how one could indeed be both. I wish we could have recorded this marvelous flower of the moment.

that summer had been, and equally how funny, how human its
events. It was an evening of laughter and tears, nostalgia and
sobriety, as we looked at each other and realized that twenty
years had passed and that almost all of those extraordinary
patients were now dead.

All, that is, save one—Lillian Tighe, who showed such eloquence in the documentary film. Bob, Robin, Penny, and I visited her, and we all marveled at her toughness, her humor, her
lack of self-pity, her realness. She had retained, despite advanc-
ing disease and unpredictable reactions to L-dopa, all of her
humor, her love of life, her spunkiness.

I spent a great deal of time on the set of *Awakenings* during
the months of filming. I showed the actors how parkinsonian
patients sat, immobile, with masked face and unblinking eyes;
the head perhaps pulled backwards or torqued to one side; the
mouth tending to hang open, a little spittle perhaps hanging
from the lips (drooling was felt to be difficult, and perhaps too
ugly, for the film, so we did not insist on this). I showed them
common dystonic postures of hands and feet; I demonstrated
tremors and tics.

I showed the actors how parkinsonian patients stood, or
tried to stand; how they walked, often bent over, sometimes
accelerating and festinating; how they might come to a halt,
freeze, and be unable to go on. I showed them different sorts of
parkinsonian voices and noises and parkinsonian handwriting.
I counseled them to imagine themselves locked in small spaces
or stuck in a vat of glue.

We practiced *kinesia paradoxa*—the sudden release from
parkinsonism by music or by spontaneous responses such as
catching a ball (the actors loved practicing this with Robin,

who we felt might make a great ballplayer were he not committed to acting). We practiced catatonia and postencephalitic card games: four patients would sit completely frozen, clutching hands of cards, until someone (perhaps a nurse) made a first move, which precipitated a tremendous flurry of movement; the game, first paralyzed, now finished itself within seconds (I had seen and captured just such a card game on film in 1969). The nearest thing to these accelerated, convulsive states is Tourette's syndrome, so I brought several young people with Tourette's to the set. These almost Zen-like exercises—becoming immobile, emptying oneself, or accelerating oneself, perhaps for hours on end—were both fascinating and frightening to the actors. They started to feel with frightful vividness what it might really be like to be stuck in this way permanently.

Can an actor with a normally functioning nervous system and physiology truly "become" someone with a profoundly abnormal nervous system, experience, and behavior? On one occasion, Bob and Robin were depicting a scene in which the doctor is testing the patient's postural reflexes (which can be absent or severely impaired in parkinsonism). I took Robin's place for a moment to show how one tests these: one stands behind the patient and, very lightly, pulls him backwards (a normal person would accommodate to this, but a parkinsonian or postencephalitic might fall backwards like a ninepin). As I demonstrated this on Bob, he fell backwards onto me, completely inert and passive, with no hint of any reflexive reaction. Startled, I pushed him gently forward to the upright position, but now he started to topple forward; I could not balance him. I had a sense of bewilderment mixed with panic. For a moment,

I thought that there had suddenly been a neurological catastrophe, that he had actually lost all his postural reflexes. Could acting like this, I wondered, actually alter the nervous system?

The next day I was talking with him in his dressing room before the day's shooting began, and as we talked, I noticed that his right foot was turned in with precisely the dystonic curvature it was held in when he portrayed Leonard L. on the set. I commented on this, and Bob seemed rather startled. "I didn't realize," he said. "I guess it's unconscious." He sometimes stayed in character for hours or days; he would make comments at dinner which belonged to Leonard, not himself, as if residues of the Leonard mind and character were still adhering to him.

By February of 1990, we were exhausted: there had been four months of filming, to say nothing of the months of research that preceded this. But one event galvanized us all: Lillian Tighe, the last surviving postencephalitic from Beth Abraham, came to visit the set, where she would play herself in a scene with Bob. What would she think of the make-believe postencephalitics around her? Would the actors pass muster? There was a feeling of awe on the set as she entered; everyone recognized her from the documentary.

I wrote in my journal that night:

However much the actors immerse themselves, identify, they are merely playing the part of a patient; Lillian has to be one for the rest of her life. They can slip out of their roles, she cannot. How does she feel about this? (How do I feel about Robin playing me? A temporary role for him, but lifelong for me.)

As Bob is wheeled in and takes up the frozen, dystonic posture of Leonard L., Lillian T., herself frozen, cocks an alert and critical eye. How does Bob, acting frozen, feel about Lillian, scarcely a yard away, actually so? And how does she, actually so, feel about him, acting so? She has just given me a wink, and a barely perceptible thumbs-up sign, meaning, "He's okay—he's *got* it! He really knows what it's like."

Voyages

At one time, my father had thought of a career in neurology but then decided that general practice would be "more real," "more fun," because it would bring him into deeper contact with people and their lives.

This intense human interest he preserved to the last: when he reached the age of ninety, David and I entreated him to retire—or, at least, to stop his house calls. He replied that home visits were "the heart" of medical practice and that he would sooner stop anything else. From the age of ninety to almost ninety-four, he would charter a mini-cab for the day to continue his house calls.

There were families he had treated for several generations, and he sometimes startled a young patient by saying, "Your great-grandfather had a very similar problem in 1919." He knew the human, the inward side of his patients no less than their bodies and felt he could not treat one without the other. (Indeed, it was often remarked that he knew the insides of his patients' refrigerators as well as the insides of their bodies.)

He would often become a friend, as well as a physician, to his patients. This intense interest in the entire lives of his patients made him, as it made my mother, a marvelous storyteller. His medical stories enchanted us as children and played a part in making Marcus, David, and me follow our parents into medicine.

Pop also had a deep and lifelong passion for music. He was

an inveterate concertgoer throughout his life, especially fond of the Wigmore Hall; he had first been taken there as a youth (when it was still called the Bechstein Hall). He would go to two or three concerts a week, right up to the last months of his life. He had been going to the Wigmore Hall for longer than anyone could remember and in his later years became as legendary, in his way, as some of the performers.

Michael, at forty-five, became closer to Pop after our mother's death and would sometimes go out with him to concerts, which he had never done before. Pop became increasingly arthritic as he entered his eighties and was glad to have Michael as a companion, and Michael, perhaps, found it easier to help an aging, arthritic parent, rather than feeling himself, as he must have done so often in the past, the sick, dependent son-patient of a father-physician.

For the next ten years, Michael led a relatively stable (although one could scarcely call it happy) existence. A level of tranquilizer was found which kept his psychoses at bay but did not have too many adverse effects. He continued to work as a messenger (of both mundane and, he again felt, arcane communications); he again enjoyed striding about London (though *The Daily Worker*, and "all that," as he put it, were now things of the past). Michael was all too conscious of his condition, and when he was in his grimmest moods, he would say, "I am a doomed man," though there was a hint of the messianic in this too: he was "doomed" as all messiahs are doomed. (When my friend Ren Weschler visited him once and asked how he was, Michael replied, "I am in Little Ease." Ren looked baffled, and Michael had to explain that Little Ease was a cell in the Tower of London so small that a man could neither stand up nor lie down in it, could never find any ease.)

But doomed or privileged, Michael felt deepening loneliness after our mother's death; our large house was now empty save for him and Pop, empty even of patients (Pop had moved his consulting rooms out of the house). Michael had never had any friends, and his relations with colleagues, even those of many decades, were courteous but never warm. His chief love was our boxer, Butch, but Butch was getting old and arthritic and could no longer keep up with Michael.

In 1984, the founder of the firm where Michael had worked for almost thirty-five years retired, and the company was sold to a larger firm, which promptly fired all the old employees. At fifty-six, Michael found himself out of work. He struggled to gain useful skills; he worked hard at learning typing, short-hand, and bookkeeping but found these traditional skills less and less valued in a rapidly changing world. Overcoming his awkwardness—he had never approached anyone for a job previously—he went for two or three interviews and was turned down. At this point, I think, he gave up hope of further work. He gave up his long walks and turned to heavy ciga-rette smoking: he spent hours sitting in the lounge, smoking and staring into space; this is how I would usually find him when I visited London in the mid- and later 1980s. For the first time in his life, at least the first time that he acknowledged, he started hearing voices. These "deejays" (he pronounced this "Die-Jays"), he told me, using some sorts of preternatural radio waves, were able to monitor his thoughts, broadcast them, and insinuate thoughts of their own.

At this point, Michael said he wanted a general doctor of his own, not our father, who had always acted as his physi-cian. Seeing that Michael looked underweight and pale and not merely "decompensated," his new physician did some simple medical tests and found that Michael had anemia and hypothyroidism. Once thyroxine, iron, and vitamin B_{12} were prescribed, Michael regained much of his energy, and in three months the "deejays" were gone.

•

In 1990, Pop died; he was ninety-four. Though David and his family in London had been a great support to Michael and Pop in his later years, we all felt it would be impossible for Michael to live alone in the big house at 37 Mapesbury or even in an apartment of his own. After a long search, we settled on a residence designed for elderly Jewish people with mental illness just up the road, at 7 Mapesbury. Here, we thought, Michael, now back in good physical health, would have a supportive structure, would know the neighborhood, and could easily walk to the synagogue, the bank, or familiar shops.

David and Lili would have Michael over for a Shabbat dinner on Friday evenings. Liz, my niece, would visit him regularly and check on all his needs. Michael agreed to all this with as good a grace as he could muster and would later joke about his relocation, saying that in his seventy-odd years the only journey he had ever made was from number 37 to number 7 Mapesbury Road. (Michael's bond with Liz was now the closest family bond in his life. She could draw him out of his grim obsessions for a while, and they would sometimes laugh and joke together.)

The residence, Ealon House, worked out surprisingly well; it gave Michael something of a social life and some practical skills. When I visited him, he would make a cup of tea or coffee for me in his room; he had never even made his own tea or coffee before. He showed me the washing machine and dryer in the basement; he had never taken care of his own laundry before, and now he not only did his own but helped older residents do theirs, too. And he began, by degrees, to assume a certain status, certain roles, in this small community.

Though he had virtually ceased reading now ("*Don't* send me any more books!" he wrote to me on one occasion), he had retained the fruits of a lifetime's reading and became a virtual encyclopedia whom the other residents could consult. Michael, who had felt ignored or taken for granted for much of his life, enjoyed his new status as a man of knowledge, a wise elder.

And after a lifetime of distrusting doctors, he came to trust the exceptional doctor, Cecil Helman, who looked after him and the other residents.[1] Cecil and I enjoyed a correspondence and then a friendship, and he often wrote to me about Michael. In one letter, he wrote:

Michael is in *good* shape at present. The staff describe his situation as "brilliant." He does the Kiddush every Friday night at Ealon House, and apparently is very good at it. It's given him an almost rabbinical role within that little community, and I believe it's helped his self-esteem a great deal.

("I have a Holy Mission, *I think*," Michael wrote to me. The capitalization of "Holy Mission" and the carefully underlined "I think" showed a sense of irony or humorous reservation about himself.)

---

1. Cecil Helman, who came from a family of rabbis and physicians, was also a medical anthropologist well known for his cross-cultural studies of narrative, medicine, and illness in South Africa and Brazil. Deeply thoughtful and a wonderful teacher, he recounted his medical training in South Africa under apartheid in his memoir *Suburban Shaman*.

When David died of lung cancer in 1992, Michael was deeply distressed. "*I* should have died!" he said, and for the first time in his life made a suicidal gesture, drinking a whole bottle of strong codeine cough syrup. (He had a very long sleep, but nothing worse.)

Otherwise, the last fifteen years of his life were relatively tranquil. He helped others and had a role, an identity he had never had at home, and he had a little life outside Ealon House, going for walks in the neighborhood and eating at a diner in Willesden Green (he enjoyed ham and eggs for dinner instead of the bland kosher food served at Ealon House). Lili and Liz, David's wife and daughter, continued to have him over on Friday evenings. I would stay in a nearby hotel when I went to London, now that the house was sold, and invite Michael there for a Sunday brunch. And a couple of times, Michael invited me to *his* diner, playing the host and paying the bill; this clearly gave him great pleasure.

When I visited him, he would always ask me to bring him a smoked salmon sandwich and a carton of cigarettes. I was happy to bring him a sandwich—smoked salmon was also my favorite food—less happy about the cigarettes; he was now chain-smoking close to a hundred cigarettes a day (their cost consumed almost his entire allowance).[2]

This heavy cigarette smoking affected Michael's health, giv-

---

2. Many of the residents at Ealon House were chain-smokers (as are many "chronic" schizophrenic patients generally). I do not know whether they smoke from boredom—there was not that much to do at the residence—or for the pharmacological effects of nicotine, whether these are rousing or calming. I once saw a patient at Bronx State who was apathetic and withdrawn for the most part but would become first animated and then hyperactive, boisterous, almost Tourettic, after a few puffs of a cigarette. The attendant called him "a nicotine Jekyll and Hyde."

ing him not only a smoker's cough and bronchitis but, more seriously, aneurysms in many of the arteries in his legs. In 2002, one of his popliteal arteries became blocked, almost cutting off blood flow to the lower part of his leg, which started to become cold and pale and no doubt painful; ischemic pain can be very severe. Michael, however, made no complaints, and it was only when he was observed limping that he was sent to a doctor. Fortunately, surgeons were able to save his leg.

Though Michael would say, "I am a doomed man!" announcing this to all and sundry in a loud, booming voice, he showed little emotion in ordinary social encounters. There was one occasion, however, when his severe demeanor softened. Our nephew Jonathan once visited Michael with his ten-year-old twin sons, and they both leapt up on this great-uncle they had never seen, showering him with endearments and kisses. Michael first stiffened, then softened, then burst into a roar of laughter, embracing his great-nephews with a warmth and spontaneity he had not shown (or perhaps felt) for years. This was immensely moving to Jonathan, who, born in the 1950s, had never seen Michael "normal."

In 2006, an aneurysm in Michael's other leg blocked up, and here again he made no complaints, though he was well aware of the dangers. He had been getting more disabled generally and knew that should he lose the leg or become more bronchitic, Ealon House would no longer be able to care for him. If this happened, he would have to be moved to a nursing home, where he would have no autonomy, identity, or role. Life under these circumstances, he felt, would be meaningless, intolerable. I wonder, then, if he willed himself to die.

The last scene of Michael's life was played out in a hospital emergency room, waiting for the operation which this time, he thought, would probably take his leg. He was lying on a stretcher when he suddenly raised himself up on an elbow, said, "I'm going outside to have a smoke," and fell back dead.

L ate in 1987, I met Stephen Wiltshire, an autistic boy in England. I was astounded by the hugely detailed architectural drawings which he had started doing at the age of six; he had only to glance at a complex building or even a whole cityscape for a few seconds before drawing the whole thing accurately from memory. Now thirteen, he had already published a book of his drawings, even though he was still withdrawn and virtually mute.

I wondered what lay beneath Stephen's extraordinary skill in instantly "recording" a visual scene and reproducing it in minute detail; I wondered how his mind worked, how he saw the world. Above all, I wondered about his capacities for emotion and for relationship with other people. Autistic people had been seen, classically, as being intensely alone, incapable of relationships with others, incapable of perceiving others' feelings or perspectives, incapable of humor, playfulness, spontaneity, creativity—mere "intelligent automata," in Hans Asperger's terms. But even my brief glimpse of Stephen had given me a much warmer impression.

Over the next couple of years, I spent a good deal of time with Stephen and his teacher and mentor, Margaret Hewson. Stephen's drawings had been published to great acclaim, and he embarked on a number of trips to draw buildings around the world. Together, we went to Amsterdam, Moscow, California, and Arizona.

I met with a number of experts on autism, including Uta

Frith in London. We talked mostly about Stephen and other savants, but as I left, she suggested that I meet Temple Grandin, a gifted scientist with a high-functioning form of autism that was just beginning in those days to be called Asperger's syndrome. Temple, she said, was brilliant and quite different from the autistic children I had met in hospitals and clinics; she had a Ph.D. in animal behavior and had written an autobiography.[3] It was becoming increasingly clear, Frith said, that autism did not necessarily mean severely impaired intelligence and an inability to communicate. There were some autistic people who might have developmental delays and a certain inability to read social cues, but they were fully capable and perhaps even highly gifted in many other ways.

I arranged to spend a weekend with Temple at her home in Colorado. I thought this might make an interesting footnote to the piece I was writing on Stephen.

Temple was at pains to be courteous, but it was clear in many ways that she had little understanding of what might be going on in other people's minds. She herself, she emphasized, did not think in linguistic terms but in very concrete, visual terms. She had great empathy for cattle, and she thought of herself as "seeing from a cow's point of view." This, combined with her brilliance as an engineer, had led her to become a world-renowned expert on designing more humane facilities for cattle and other animals. I was very moved by her obvious intelligence and her yearning to communicate, so different

---

3. Temple's first book, *Emergence: Labeled Autistic*, was published in 1986, a time when Asperger's syndrome was barely recognized. In it, she spoke of her "recovery" from autism; at that time, it was generally felt that no one with autism could go on to lead a productive life. By the time I met her in 1993, Temple was no longer speaking of "curing" autism but of the strengths and weaknesses people with autism may show.

from Stephen's passivity and seeming indifference to others. As she hugged me good-bye, I knew that I would have to write a long essay about her.

A couple of weeks after I had sent my piece about Temple to *The New Yorker*, I happened to see Tina Brown, the new editor of the magazine, and she said to me, "Temple will be an American hero." She proved to be right. Temple is a hero now to many in the autism community around the world, widely admired for forcing all of us to see autism and Asperger's not as neurological deficits so much as different modes of being, ones with their own unique dispositions and needs.

My earlier books had shown patients struggling to survive and adapt (often ingeniously) to various neurological conditions or "deficits," but for Temple and many of the others I wrote about in *An Anthropologist on Mars*, their "conditions" were fundamental to their lives and often a source of originality or creativity. I subtitled the book "Seven Paradoxical Tales," because all of its subjects had found or created unexpected adaptations to their disorders; all had compensating gifts of different sorts.

•

In 1991, I got a phone call about a man (I called him Virgil when I wrote about him in *An Anthropologist on Mars*) who had been virtually blind since early childhood from retinal damage and cataracts. Now, at the age of fifty, he was about to get married, and his fiancée had urged him to have cataract surgery—what was there to lose? She hoped he might begin a new life as a sighted man.

But when the bandages were removed after surgery, no miraculous cry ("I can see!") burst from Virgil's lips. He seemed to be staring blankly, bewildered, without focusing on the surgeon who stood before him. Only when the surgeon spoke—saying, "Well?"—did a look of recognition cross Virgil's face. He knew that voices came from faces, and he deduced that the chaos of light and shadow and movement he saw must be the face of his surgeon.

Virgil's experience was almost identical with that of SB, a patient whom the psychologist Richard Gregory had described thirty years earlier, and I spent many hours discussing Virgil's case with him.

Richard and I had met in Colin Haycraft's office back in 1972, when Colin was preparing to publish not only *Awakenings* but Gregory's book *Illusion in Nature and Art*. He was a large man, a head taller than I, with a spontaneity, exuberance, and energy of mind and body, combined with a sort of innocence and a fondness for jokes that made me think of him as a huge, ebullient, jocular boy of twelve. I had been enchanted by his earlier books—*Eye and Brain* and *The Intelligent Eye*, which showed the easy, delicious working of a powerful, passionate mind and a characteristic amalgam of playfulness and profundity. One could recognize a Gregory sentence as easily as a bar of Brahms.

We both had a special interest in the brain's visual system and how our powers of visual recognition could be undermined by injury or disease or tricked by visual illusions.[4] He felt strongly that perceptions were not just simple reproductions of sensory data from the eye or ear but had to be "constructed" by

---

4. Many generations of Gregorys had been specially interested in vision and optics. In his book *Hereditary Genius*, Francis Galton traced the intellectual eminence of the Gregory family back to Newton's contemporary James Gregory, who made important improvements to Newton's reflector telescope. Richard's own father had been the astronomer royal.

the brain, a construction involving the collaboration of many subsystems in the brain and constantly informed by memory, probability, and expectations.

Over a long and productive career, Richard showed that visual illusions provided a major way of understanding all sorts of neurological functions. Play was central to him, both intellectual play (he was an avid punster) and as a method in science. He felt that the brain played with ideas, that what we called perceptions were really "perceptual hypotheses" that the brain constructed and played with.

When I lived on City Island, I often got up in the middle of the night to ride my bicycle when the streets were empty, and one night I noticed an odd phenomenon: if I looked at the spokes of the front wheel as it was revolving, there might be a moment when they appeared frozen, as in a still photograph. This fascinated me, and I instantly phoned Richard, forgetting that it must be very early in the morning for him in England. But he took this cheerfully and presented me with three on-the-spot hypotheses. Was the "frozenness" a stroboscopic effect caused by the oscillating current from my dynamo? Was it due to the jerking, saccadic movements of my eyes? Or did it indicate that the brain in fact "constructed" a sense of motion from a series of "stills"?[5]

We both had a passion for stereoscopic vision; Richard would sometimes send stereo Christmas cards to his friends, and his museum-like house in Bristol was full of old stereoscopes, along with other old optical instruments of every sort. I consulted him often when I was writing about Susan Barry ("Ste-

---

5. I later discussed such "snapshot" vision with Francis Crick and wrote about it in "In the River of Consciousness," a 2004 essay for *The New York Review of Books*.

reo Sue"), who while apparently stereo-blind from ea
had nonetheless acquired stereo vision at the age of fift
was an achievement that was considered impossible, current
opinion being that there was only a short critical period for stereo experience in early childhood and that if stereo vision was
not achieved by the age of two or three, it would be too late.

And then, in the wake of Stereo Sue, I started to lose some
of my own vision in one eye, eventually losing it entirely. I
would write to Richard about the sometimes frightening things
which were happening to my vision and how, after a lifetime
of seeing the world in rich, beautiful stereo depth, I now found
it so flat and confusing that I seemed at times to lose the very
concepts of distance and depth. Richard was endlessly patient
with my questions, and his insights were invaluable. I felt that
he, more than anyone, helped me make sense of what I was
experiencing.

Early in 1993, Kate handed me the phone and said, "It's John
Steele, calling from Guam."

Guam? I had never had a phone call from Guam. I was
not even sure where it was. I had had a little correspondence
twenty years earlier with a John Steele, a neurologist in
Toronto who had co-authored an article on migraine hallucinations in children. That John Steele was known for identifying
Steele-Richardson-Olszewski syndrome, a degenerative brain
disease now called progressive supranuclear palsy. I lifted the
phone, and it turned out to be the same John Steele. He told me
how he had since made his life in Micronesia, first in some of
the Caroline Islands and now in Guam. Why was he phoning
me? He said there was an extraordinary disease called lytico-

bodig endemic among the native Chamorro people in Guam. Many of them had symptoms extremely similar to what I had described and filmed in my postencephalitic patients. Because I was among the very few people now who had seen such postencephalitics, John wondered if I could meet some of his patients and give him my thoughts.

I remembered hearing about the Guam disease as a resident; it was sometimes considered the Rosetta stone of neurodegenerative diseases, for patients with it often showed symptoms like those of Parkinson's or ALS or dementia and could perhaps cast light on all of these. Neurologists had gone to Guam for decades trying to crack the cause of the disease, but most of them had given up.

I arrived in Guam a few weeks later, and I was met at the airport by John, an instantly recognizable figure. It was swelteringly hot, and everyone was in colorful shirts and shorts except John, who was nattily dressed in a tropical suit, a tie, and a straw hat. "Oliver!" he shouted. "So good of you to come!"

As he drove us away in his red convertible, he filled me in on the history of Guam; he also pointed out stands of cycads, a very primitive tree which had originally forested the whole of Guam; he knew that I was interested in cycads and other primitive plant forms. Indeed, he had suggested on the phone that I could come to Guam either as a "cycadological neurologist" or as a "neurological cycadologist," for many people thought that a flour made from the seeds of these cycads, a popular Chamorro food, was responsible for the strange disease there.

For the next few days, I went with John on his house calls. It reminded me of how, as a boy, I used to go with my father on house calls. I met many of John's patients, and some of them did indeed remind me of my *Awakenings* patients. I decided I wanted to return to Guam for a longer visit—this time with a camera, to film some of these unique patients.

I found the Guam visit very important at a human level, too. While the postencephalitic patients had been put away for decades, living in a hospital, often abandoned by their families, people with lytico-bodig remained part of their family, part of their community, to the end. This drove home to me how barbaric our own medicine and our own customs are in the "civilized" world, where we put ill or demented people away and try to forget them.

·

One day in Guam, I got talking to John about another interest of mine, color blindness, a subject I had been deeply interested in for years. I had recently seen a painter, Mr. I., who had suddenly lost the ability to perceive color after a lifetime of seeing in color. He knew what he was missing, but if one were *born* without the ability to see color, then one would have no idea of what color was like. Most "colorblind" people are really color deficient: they have difficulties discriminating certain colors but can easily see others. But the inability to see any color, total congenital color blindness, is extremely rare; it affects perhaps one person in thirty thousand. How would people with such a condition do in a world that, for other people, and for birds and mammals, is full of informative and suggestive colors? Might such achromatopes, like deaf people, develop special, compensatory skills and strategies? Might they create, like the deaf, a whole community and culture?

I mentioned to John that I had heard a rumor—a romantic legend, perhaps—about an isolated valley populated entirely by people who were totally colorblind. John said, "Yes, I know the place. It's not exactly a valley, but it is very isolated, a tiny coral atoll relatively close to Guam—only twelve hundred miles away." The island, Pingelap, was near Pohnpei, a larger volcanic island where John had worked for some years. He said he had seen some Pingelapese patients on Pohnpei, and he understood that about 10 percent of Pingelap's population was totally colorblind.

·

A few months later, Chris Rawlence, who had written the libretto for Michael Nyman's opera of *The Man Who Mistook His Wife for a Hat*, proposed doing a series of documentaries with me for the BBC.[6] And so we returned to Micronesia in 1994, accompanied by my ophthalmologist friend Bob Wasserman, and Knut Nordby, a Norwegian psychologist who was himself totally colorblind. Chris and his crew organized a precariously small plane to get us to Pingelap, and Bob, Knut, and I immersed ourselves in the unique cultural life and history of these islands. We saw patients and spoke to doctors, botanists, and scientists; we wandered in rain forests, we snorkeled in the reefs, we sampled the intoxicating *sakau*.

It was not until the summer of 1995 that I settled down to write about these island experiences, and I really conceived of the book as a pair of narrative travelogues: "The Island of the Colorblind," about Pingelap; conjoined with "Cycad Island," about the strange disease on Guam. (To these I added a sort of coda about deep, geological time and my favorite ancient plants, cycads.)

------

6. The series, called *The Mind Traveller*, explored a number of subjects I had long been interested in, including Tourette's syndrome and autism. It also introduced me to some new experiences—with people with Williams syndrome (whom I would later write about in *Musicophilia*), with a deaf-blind Cajun community, and with a number of deaf languageless people.

I felt free to explore many non-neurological topics as well as neurological ones, and it included more than sixty pages of endnotes, many of them little essays about botany or mathematics or history. So *Island* was different from any of my previous books: more lyrical, more personal. It remains, in some ways, my favorite book.

Nineteen ninety-three not only saw the beginning of new adventures and travels in Micronesia and elsewhere but launched me on another journey, one of mental time travel, recollecting and revisiting in memory some of the passions of my own early years.

Bob Silvers asked me if I would review a new biography of Humphry Davy. This thrilled me, for Davy was an idol of mine when I was a boy; I loved reading about his chemical experiments in the early nineteenth century and repeating them in my little lab. I reimmersed myself in the history of chemistry, and I got to know the chemist Roald Hoffmann.

A few years later, Roald, knowing of my boyhood passion for chemistry, sent me a parcel containing a large poster of the periodic table with photographs of each element, a chemical catalogue, and a little bar of a very dense, greyish metal which I immediately recognized as tungsten. As Roald no doubt surmised, this instantly provoked memories of my uncle whose factory had produced bars of tungsten and manufactured lightbulbs with tungsten filaments. That bar of tungsten was my madeleine.

I began writing about my boyhood, growing up in England before the Second World War, being exiled to a sadistic boarding school during the war, and finding constancy in my passion for numbers and later for elements and the beauty of equations which could represent any chemical reaction. It was a new sort of book for me, combining memoir with a sort of history of chemistry. By the end of 1999, I had written many hundreds of thousands of words, but I felt the book did not quite come together.

•

I used to delight in the natural history journals of the nineteenth century, all of them blends of the personal and the scientific—especially Wallace's *Malay Archipelago,* Bates's *Naturalist on the River Amazons,* and Spruce's *Notes of a Botanist,* and the work which inspired them all (and Darwin too), Alexander von Humboldt's *Personal Narrative.* It pleased me to think that Wallace, Bates, and Spruce were all crisscrossing one another's paths, leapfrogging, on the same stretch of the Amazon during the selfsame months of 1849 and to think that all of them were good friends. (They continued to correspond throughout their lives, and Wallace was to publish Spruce's *Notes* after his death.)

They were all, in a sense, amateurs—self-educated, self-motivated, not part of an institution—and they lived, it sometimes seemed to me, in a halcyon world, a sort of Eden, not yet turbulent and troubled by the almost murderous rivalries which were soon to mark an increasingly professionalized world (the sorts of rivalries so vividly portrayed in H. G. Wells's story "The Moth").

This sweet, unspoiled, preprofessional atmosphere, ruled by a sense of adventure and wonder rather than by egoism and a lust for priority and fame, still survives here and there, it seems to me, in certain natural history societies, whose quiet yet essential existences are virtually unknown to the public.

One such is the American Fern Society, which holds monthly meetings and occasional field trips—"fern forays"—of one sort or another.

In January of 2000, still wrestling with how to complete *Uncle Tungsten,* I took a trip with about twenty members of the fern society to Oaxaca, where more than seven hundred species of fern have been described. I had not planned to keep a detailed journal, but there was such a sense of adventure and richness of experience that I wrote almost nonstop for the entire ten-day trip.[7]

The block I had felt with *Uncle Tungsten* broke suddenly in the middle of Oaxaca city as I boarded a shuttle bus in the town square to go back to my hotel. Sitting opposite me on the bus were a cigar-smoking man and his wife, both speaking Swiss German. The conjunction of the shuttle bus and the language took me back suddenly to 1946, as I wrote in *Oaxaca Journal:*

The war had just ended, and my parents decided to visit Europe's only "unspoiled" country, Switzerland. The Schweizerhof in Lucerne had a tall, silent electric brougham which had been running quietly and beautifully since it was made, forty years earlier. A sudden half-sweet, half-painful memory comes up of my thirteen-year-old self on the verge of adolescence. The freshness and sharpness of all my perceptions then. And my parents—young, vigorous, just fifty.

---

7. When I returned, I transcribed the journal, and shortly thereafter I was invited to publish it as a book in a National Geographic travel series. There are whole pages of the published *Oaxaca Journal* that are identical with the handwritten journal, but I also enriched it with research on other things which had struck me during the trip—chocolates and chilis, mescal and cochineal, Mesoamerican culture and New World hallucinogens.

When I returned to New York, memories of boyhood continued to come to me, and the rest of *Uncle Tungsten* followed,
the personal seeming to weave itself into the historical and the
chemical—so this hybrid of a book came into being, with two
very different stories and voices somehow meshed together.

S omeone who shared my deep love of natural history and the
history of science was Stephen Jay Gould.

I had read his *Ontogeny and Phylogeny* and many of his
monthly articles in *Natural History* magazine. I particularly
liked his 1989 book *Wonderful Life*, which gave one a tremendous feeling for the sheer luck—good or bad—which can befall
any species of animal or plant and the huge role that chance
plays in evolution. As he wrote, if we could "rerun" evolution, it would no doubt turn out completely differently every
time. *Homo sapiens* was the result of a particular combination
of contingencies that ended up producing us. He called this a
"glorious accident."

I was so excited by Gould's vision of evolution that when
asked by a newspaper in England what book I had most enjoyed
in 1990, I selected *Wonderful Life*, his vivid evocation of the
astonishing range of life-forms produced in the "Cambrian
explosion" more than 500 million years ago (these had been
beautifully preserved in the Burgess Shale in the Canadian
Rockies) and how many of these had succumbed to competition, disaster, or just bad luck.

Steve saw this tiny book review and sent me a generously
inscribed copy of the book, in which he spoke of it as "the geological version" of the sort of contingency, the inherent unpre-

dictability, I had described in my postencephalitic patients. I thanked him, and he replied with a letter crackling with his special energy, exuberance, and style. It began:

Dear Dr. Sacks,

I was thrilled to get your letter. There can hardly be a greater pleasure in life than learning that an intellectual hero has enjoyed one's own labor in return. I really do think that, in some collective sense but obviously without any contact, several of us are working towards a common goal rooted in a theory of contingency. Your work on case studies certainly goes together with Edelman on neurology, chaos theory in general, McPherson on the Civil War, and my own material on the history of life. There is, of course, nothing new about contingency *per se*. Rather, the theme has usually been seen either as something outside science ("merely history") or, even worse, as a surrogate, or even a rallying point, for an unscientific spiritualism. The point is not to stress contingency, but to identify it as a central theme for a genuine science based on the irreducibility of individuality, not as something standing against science but as an expectation of what we call natural law, and therefore as a primary datum of science itself.

After discussing several other topics, he concluded:

Funny how once you get in contact with someone you wanted to meet for years, you begin to see things you want to discuss with him everywhere.

Sincerely,

Stephen Jay Gould

We did not, in fact, meet until a couple of years later, when a television journalist in Holland approached us to do a series of interviews. When the producer asked if I knew Steve, I had replied, "I've never met him, although we've corresponded. But nonetheless, I think of him as a brother."

Steve, for his part, had written to the producer, "I desperately want to meet Oliver Sacks. I see him as a brother, but we've never met."

There were six of us in all—Freeman Dyson, Stephen Toulmin, Daniel Dennett, Rupert Sheldrake, Steve, and myself. We were each interviewed separately and then, a few months later, flown to Amsterdam, where we were put in separate hotels. None of us had yet met the others, and there was a hope that there would be some wonderful (and possibly violent) explosion as the six of us came together. The thirteen-hour television show, called *A Glorious Accident*, was a huge hit in the Netherlands, and a transcript of the show became a best-selling book.

Steve's own response to the show was characteristically puckish. He wrote, "I am astonished to see that our Dutch series was so well received. I certainly enjoyed meeting you all immensely, but I doubt that I would have been inclined to spend hours before a TV set watching such a conversation among a group of folks usually characterized in these p.c. days as dead white European males."

Steve taught at Harvard, but he lived in downtown New York, so we were neighbors. There were so many different aspects to Steve, so many passions. He loved walking, and he had a huge architectural knowledge of New York City and what it looked like a century ago. (Only someone as intensely sensitive to architecture as he would introduce spandrels as an evolutionary metaphor.) He was extremely musical: he sang in a choir in Boston, and he adored Gilbert and Sullivan; I think he knew all of Gilbert and Sullivan by heart. On one occasion when we went out to visit a friend on Long Island, Steve basked in the hot tub for three hours, all the while singing Gilbert and Sullivan and never repeating himself. He also knew a huge number of songs from both world wars.

Steve and his wife, Rhonda, were impulsively generous friends, and they loved to host birthday parties. Steve would bake a birthday cake using his mother's recipe, and he would always write a poem to recite. He was very good at this; one year he turned out a marvelous version of "Jabberwocky," and at another party he recited this:

FOR OLIVER'S BIRTHDAY, 1997

This man, who's in love with a cycad
But once could have starred in a bike ad
King of multidiversity
Hip! Happy birth-i-day
You exceed what old Freud, past head psych, had.

One legg'd, migrained, color blinded
Awak'ning on Mars, and hat-minded
Oliver Sacks
Still lives life to the max
While his swimming leaves dolphins behinded.

On another birthday, knowing that I loved the periodic table, Steve and Rhonda invited everyone to dress as a particular element. I am rather bad at names and faces, but I never forget an element. (There was one man who came to the party with my old friend Carol Burnett. I do not remember his name, and I cannot remember his face, but I will always remember him as argon.) Steve was xenon, element 54, another noble gas.

•

I eagerly read Steve's monthly articles in *Natural History* and often wrote to him about subjects he raised. We discussed all sorts of things, from the place of contingency in the reactions of patients to our shared love for museums (especially the old cabinet type; we both spoke out for the preservation of the marvelous Mütter Museum in Philadelphia).

I also had a craving which went back to my marine biology days to know more about more primitive nervous systems and behaviors, and here Steve was an important influence in my life, someone who reminded me, incessantly, that nothing in biology made sense except in the light of evolution and chance, contingency. He put everything in the context of deep evolutionary time.

Steve's own research had been on the evolution of land snails in Bermuda and in the Netherlands Antilles, and for him the vast range of invertebrates illustrated even better than vertebrates the range of nature's inventiveness and its ingenuity in finding new uses for very early evolved structures and mechanisms of every sort—he called these "exaptations." So we shared an appreciation for "lower" life-forms.

In 1993, I wrote to Steve of ways of joining particulars with generalities—in my own case, clinical narratives with neuroscience—and he replied, "I have long experienced exactly the same tension, trying to assuage my delight in individual things through my essays and my interest in generality through my more technical writing. I loved the Burgess Shale work so much because it allowed me to integrate the two."

He was kind enough to read my manuscript for *The Island of the Colorblind,* and he did so closely, saving me from a number of blunders.

Finally, we had in common an interest in autism; as he wrote to me, "My reasons for respect are partly personal. I have an autistic son, who is one of the great day/date calculators—instantaneously, over thousands of years. Your piece on the calculating twins is the most moving essay I have ever read."

He had written very movingly about Jesse, his son, in an essay later published in *Questioning the Millennium:*

> Humans are storytelling creatures preeminently. We organize the world as a set of tales. How, then, can a person make any sense of his confusing environment if he cannot comprehend stories or surmise human intentions? In all the annals of human heroics, I find no theme more ennobling than the compensations that people struggle to discover and implement when life's misfortunes have deprived them of basic attributes of our common nature.

Steve had had a brush with death before I met him, when he was forty or so. He had a very rare malignant tumor—a mesothelioma—but was determined to beat the odds and survive this particularly lethal cancer. He was one of the lucky ones, aided by radiation and chemotherapy. He had always been an extremely energetic person, but after this experience of facing death, he became more energetic than ever. There was not a minute to waste; who knew what might happen next?

Twenty years later, at the age of sixty, he developed a seemingly unrelated cancer—a lung cancer in the chest that metastasized to the liver and the brain. But the only concession he made to illness was to sit while lecturing instead of standing. He was determined to complete his magnum opus, *The Structure of Evolutionary Theory*, and it came out in the spring of 2002, the twenty-fifth anniversary of his publication of *Ontogeny and Phylogeny*.

A few months later, just after teaching his final class at Harvard, Steve plunged into coma and died. It was as if he had kept himself going by sheer willpower, and then, having completed his final semester of teaching, having seen his final book published, he was ready to let things go. He died at home in his library, surrounded by the books he loved.

A New Vision of the Mind

Early in March of 1986, soon after *Hat* was published, I received a letter from Mr. I., an artist on Long Island. He wrote:

I am a rather successful artist just past 65 years of age. On January 2nd of this year I was driving my car and was hit by a small truck on the passenger side of my vehicle. When visiting the emergency room of a local hospital, I was told I had a concussion. While taking an eye examination, it was discovered that I was unable to distinguish letters or colors. The letters appeared to be Greek letters. My vision was such that everything appeared to me as viewing a black and white television screen. Within days, I could distinguish letters and my vision became that of an eagle—I can see a worm wriggling a block away. The sharpness of focus is incredible. BUT—I AM ABSOLUTELY COLOR BLIND. I have visited ophthalmologists who know nothing about this color-blind business. I have visited neurologists, to no avail. Under hypnosis I still can't distinguish colors. I have been involved in all kinds of tests. You name it. My brown dog is dark grey. Tomato juice is black. Color TV is a hodge-podge.

Mr. I. complained that the dreary, "leaden" black-and-white world he now inhabited made people look hideous and painting impossible. Had I encountered such a condition before? Could I track down what had happened? Could I help him?

I replied that I had heard of such cases of acquired achromatopsia but never seen one. I was not sure if I could help, but I invited Mr. I. to come and see me.

Mr. I. had became colorblind after sixty-five years of seeing colors normally—totally colorblind, as if "viewing a black and white television screen." The suddenness of the event was incompatible with any of the slow deteriorations that can befall the retinal cone cells and suggested instead a mishap at a much higher level, in those parts of the brain specialized for the perception of color.

Moreover, it became apparent that Mr. I. had lost not only the ability to see color but the ability to *imagine* it. He now dreamed in black and white, and even his migraine auras were drained of color.

A few months earlier, I had been in London for the publication of *Hat* when a colleague invited me to come along to a conference at the National Hospital in Queen Square. "Semir Zeki will be talking," he said. "He's the cat's whisker on color perception."

Zeki had been making a neurophysiological investigation of color perception by recording from electrodes inserted into the visual cortex of monkeys, and he had shown that a single area (V4) was responsible for the construction of color. He thought there was probably an analogous area in the human brain. I was fascinated by Zeki's talk, especially by his use of the word "construction" in relation to color perception.

A whole new way of thinking seemed to ray out from Zeki's work, and it set me thinking of the possible neural basis for consciousness in a way I had never considered before—and to realize that with our new powers of imaging the brain and our newly developed abilities to record the activity of individual neurons in living and conscious brains, we might be able to plot how and where all sorts of experiences are "constructed." This was an exhilarating thought. I realized the vast leap which neurophysiology had made since my own student days in the early 1950s, when it was beyond our power, almost beyond imagination, to record from individual nerve cells in the brain while an animal was conscious, perceiving, and acting.

·

Around this time, I went to a concert in Carnegie Hall. The program included Mozart's great Mass in C Minor and, after the interval, his Requiem. A young neurophysiologist, Ralph Siegel, chanced to be sitting a few rows behind me; we had seen each other briefly the previous year when I had visited the Salk Institute, where he was one of Francis Crick's protégés. When Ralph saw that I had a notebook on my lap and was writing nonstop throughout the concert, he knew the bulky figure ahead of him had to be me. He came up and introduced himself at the end of the concert, and I recognized him at once—not by his face (most faces look the same to me), but by his flaming red hair and his brash, ebullient manner.

Ralph was curious—what had I been writing about through the entire concert? Had I been wholly unconscious of the music? No, I said, I was conscious of the music, and not just as background. I quoted Nietzsche, who used to write at concerts, too; he loved Bizet and once wrote, "Bizet makes me a better philosopher."

I said I felt that Mozart made me a better neurologist and that I had been writing about a patient I had been seeing—the colorblind artist. Ralph was excited; he had heard of Mr. I., for I had described him to Francis Crick earlier in the year. Ralph's own work was exploring the visual system in monkeys, but he said he would love to meet Mr. I., who would be able to tell him exactly what he was seeing (or not seeing), unlike the monkeys he worked with. He outlined half a dozen simple but crucial tests that could help pinpoint at what stage the construction of color had broken down in the painter's brain.

•

Ralph thought always in deep physiological terms, while neurologists, myself included, often content ourselves with the phenomenology of brain disease or damage, with little thought of the precise mechanisms involved and no thought at all of the ultimate question of how experience and consciousness emerged from brain activity. For Ralph, all the questions he explored in the monkey brain, the insights he so patiently collected one by one, always pointed to that ultimate question—the relationship of brain and mind.

Whenever I told him stories about what my patients were experiencing, Ralph would immediately pull me into a physiological discussion: What parts of the brain were involved? What was going on? Could we simulate it on a computer? He was a good natural mathematician, with a degree in physics, and he enjoyed computational neuroscience, making models or simulations of neurological systems.[1]

For the next twenty years, Ralph and I were great friends.

---

1. He was fascinated when I showed him the complex patterns one might see in a migraine aura—hexagons and geometrical patterns of many shapes, including fractal patterns. He was able to simulate some of these basic patterns on a neural network, and in 1992 we included this work as an appendix to a revised edition of *Migraine*. Ralph's mathematical and physical intuition also led him to feel that chaos and self-organization might be central to natural processes of all kinds, relevant to every sort of science from quantum mechanics to neuroscience, and this led in 1990 to another collaboration between us, an appendix for a revised edition of *Awakenings*, "Chaos and Awakenings."

He spent his summers at the Salk Institute, and I often went there to visit him. As a scientist, he was uncompromising, often blunt and outspoken; as a person, he was jovial, spontaneous, and playful. He loved being a husband and father to his twins—a family life in which I was often included as a sort of godfather. We both loved La Jolla, where we could go for long walks or bicycle rides, watch the paragliders hovering over the bluffs, or swim in the cove. La Jolla had become the neuroscience capital of the world by 1995, with the Salk Institute, the Scripps Research Institute, and UCSD being joined by Gerald Edelman's Neurosciences Institute. Ralph introduced me to some of the many neuroscientists working at the Salk, and I started to feel myself part of this extraordinarily varied and original community.

In 2011, Ralph died, far too young, from brain cancer, at the age of fifty-two. I miss him deeply, but like so many of my friends' and mentors' his voice has become an integral part of my thinking.

•

In 1953, while I was at Oxford, I read Watson and Crick's famous "double helix" letter when it was published in *Nature*. I would like to say that I immediately saw its tremendous significance, but this was not the case for me, nor indeed for most people at the time.

It was only in 1962, when Crick came to San Francisco and spoke at Mount Zion Hospital, that I started to realize the vast implications of the double helix. Crick's talk was not on the configuration of DNA but on the work he had been doing with the molecular biologist Sydney Brenner to determine how the sequence of DNA bases could specify the amino acid sequence in proteins. They had just shown, after four years of intense work, that the translation involved a three-nucleotide code. This was itself a discovery no less momentous than the discovery of the double helix.

But clearly Crick had already moved on to other things. There were, he intimated in his talk, two great enterprises whose exploration lay in the future: understanding the origin and nature of life, and understanding the relation of brain and mind—in particular, the biological basis of consciousness. Did he have any inkling, when he spoke to us in 1962, that these would be the very subjects he himself would address in the years to come, once he had "dealt with" molecular biology, or at least taken it to the stage where it could be delegated to others?

In 1979, Crick published "Thinking About the Brain," an article in *Scientific American* which, in a sense, legitimated the study of consciousness in neuroscientific terms; prior to this, the question of consciousness was felt to be irretrievably subjective, and therefore inaccessible to scientific investigation.

A few years later, I met him at a 1986 conference in San Diego. There was a big crowd, full of neuroscientists, but when it was time for dinner, Crick singled me out, seized me by the shoulders, and sat me down next to him, saying, "Tell me stories!" In particular, he wanted stories of how vision might be altered by brain damage or disease.

I have no memory of what we ate, or anything else about the dinner, only that I told him stories about many of my patients and that each one set off bursts of hypotheses and suggestions for further investigation in his mind. Writing to him a few days later, I said that the experience was "a little like sitting next to an intellectual nuclear reactor. . . . I never had a feeling of such incandescence." He was fascinated when I told him about Mr. I. and also when I told him how a number of my patients had experienced, in the few minutes of a migraine aura, a flickering of static, "frozen" images in place of their normal, continuous visual perception. He asked me whether such "cinematic vision," as I called it, was ever a permanent condition or one that could be elicited in a predictable way so that it could be investigated. (I said I did not know.)

•

During 1986, I spent a good deal of time with Mr. I., and in January of 1987 I wrote to Crick, "I have now written up a longish report on my patient. . . . Only in the actual writing did I come to see how color might indeed be a (cerebro-mental) construct."

I had spent most of my professional life wedded to notions of "naive realism," regarding visual perceptions, for example, as mere transcriptions of retinal images; this "positivist" view was the dominant one in my Oxford days. But now, as I worked with Mr. I., this was giving way to a very different vision of the brain-mind, a vision of it as essentially constructive or creative. I added that I had now started to wonder whether all perceptual qualities, including the perception of motion, were similarly constructed by the brain.[2]

---

2. A few days later, I got a reply in which Crick sought more detail about the difference between my migraine patients and a remarkable patient described in a 1983 paper by Josef Zihl and his colleagues. Zihl's patient, for example, could not pour a cup of tea; she saw a motionless "glacier" of tea hanging from the spout. Some of my migraine patients had experienced such "stills" in rapid succession, whereas for Zihl's patient, who had acquired motion blindness following a stroke, the stills apparently lasted much longer, perhaps several seconds each. In particular, Crick wanted to know whether successive stills in my migraine patients occurred within the interval between successive eye movements or only between such intervals. "I would very much like to discuss these topics with you," he wrote, "including your remarks about color as a cerebro-mental construct."

Writing back to Crick, I enlarged on the deep differences between my migraine patients and Zihl's motion-blind woman.

I mentioned in my letter that I was working on Mr. I.'s case with my ophthalmologist friend Bob Wasserman and with Ralph Siegel, who had designed and conducted a variety of psychophysical experiments with our patient. I mentioned that Semir Zeki too had seen Mr. I. and tested him.

At the end of October in 1987, I was able to send Crick "The Case of the Colorblind Painter," a paper that Bob Wasserman and I had written for *The New York Review of Books*, and early in January of 1988 I got a response from Crick—an absolutely stunning letter: five pages of single-spaced typing, minutely argued and bursting with ideas and suggestions, some of which, he said, were "wild speculation." He wrote:

> Thank you so much for sending me your fascinating article on the color-blind artist. . . . Even though, as you stress in your letter, it is not strictly a scientific article, nevertheless it has aroused much interest among my colleagues and my scientific and philosophical friends here. We have had a couple of group sessions on it and in addition I have had several further conversations with individuals.

He added that he had sent a copy of the article and his letter to David Hubel, who, with Torsten Wiesel, had done pioneering work on the cortical mechanisms of visual perception. I was very excited to think that Crick was opening our paper, our "case," for discussion in this way. It gave me a deeper sense of science as a communal enterprise, of scientists as a fraternal, international community, sharing and thinking on each other's work, and of Crick himself as a sort of hub, in touch with everyone in this neuroscientific world.

"Of course the most interesting feature," Crick wrote:

is Mr. I.'s loss of the subjective sense of color, together with its absence in his eidetic imagination and in his dreams. This clearly suggests that a crucial part of the apparatus needed for these latter two phenomena is also needed for color perception. At the same time, his memory for color names and color associations remained completely intact.

He went on to carefully summarize a number of papers by Margaret Livingstone and David Hubel outlining their theory of three stages in early visual processing and speculated that Mr. I. had sustained damage at one of these levels (the "blob system" in V1), where cells would be particularly sensitive to lack of oxygen (perhaps caused by a small stroke or even carbon monoxide poisoning).

"Do please excuse the length of this letter," he concluded. "We might talk about it over the phone, after you've had time to digest it all."

Bob, Ralph, and I were all mesmerized by Crick's letter. It seemed to get deeper and more suggestive every time we read it, and we got the sense that it would need a decade or more of work to follow up on the torrent of suggestions Crick had made.

Contacting me again a few weeks later, Crick mentioned two of Antonio Damasio's cases: in one of these, the patient had lost color imagery but still dreamed in color. (She later regained her color vision.)

And he wrote:

So glad to . . . learn that you plan more work on Mr. I. All the things you mention are important, especially the scans. . . . There is no consensus yet among my friends about what the damage might be in such cases of cerebral achromatopsia. I have (very tentatively) suggested the V1 blobs plus some subsequent degeneration at higher levels, but this really depends on seeing little in the scans (if most of V4 is knocked out you should see something). David Hubel tells me that he favors damage to V4, though this opinion is preliminary. David van Essen tells me that he suspects some area further upstream.

"I think the moral of all this," Crick concluded, "is that only careful and extensive psychophysics on [such] a patient plus accurate localization of the damage will help us. (So far, we cannot see how to study visual imagery and dreams in a monkey.)"

·

In August of 1989, Crick wrote to me, "At the moment I am trying to come to grips with visual awareness, but so far it remains as baffling as ever." He enclosed the manuscript of a paper called "Towards a Neurobiological Theory of Consciousness," one of the first synoptic articles to come out of his collaboration with Christof Koch at Caltech. I felt very privileged to see this manuscript, in particular their carefully laid-out argument that an ideal way of entering this seemingly inaccessible subject would be through exploring disorders of visual perception.

Crick and Koch's paper was aimed at neuroscientists and covered a vast range in a few pages; it was sometimes dense and highly technical. But I knew that Crick could also write in a very accessible and witty and personable way; this was especially evident in his two earlier books, *Life Itself* and *Of Molecules and Men*. So I now entertained hopes that he might give a more popular and accessible form to his neurobiological theory of consciousness, enriched with clinical and everyday examples. (He did this in his 1994 book, *The Astonishing Hypothesis*.)

•

In June of 1994, Ralph and I met Crick for dinner in New York. The talk ranged in all directions. Ralph talked about his current work with visual perception in monkeys and his thoughts on the fundamental role of chaos at the neuronal level; Francis spoke about his expanding work with Christof Koch and their latest theories about the neural correlates of consciousness; and I spoke about my upcoming visit to Pingelap, with its scores of people—nearly 10 percent of the population—born completely colorblind. I planned to travel there with Bob Wasserman and Knut Nordby, a Norwegian perceptual psychologist who, like the Pingelapese, had been born without color receptors in his retinas.

In February of 1995, I sent Francis a copy of *An Anthropologist on Mars*, which had just been published and contained an expanded version of "The Case of the Colorblind Painter," much amplified, in part, through my discussions with him on the case. I also told him something of my experiences in Pingelap and how Knut and I tried to imagine what changes might have occurred in his brain in response to his achromatopsia. In the absence of any color receptors in his retinas, would the color-constructing centers in his brain have atrophied? Would they have been reallocated for other visual functions? Or were they, perhaps, still awaiting an input, an input that might be provided by direct electrical or magnetic stimulation? And if this could be done, would he, for the first time in his life, see color? Would he know it *was* color, or would this visual experience be too novel, too confounding, to categorize? Questions like these, I knew, would fascinate Francis too.

Francis and I continued to correspond on various subjects. I wrote to him at length about the patient whom I called Virgil, whose sight was restored after a lifetime of blindness, and I wrote to him with thoughts about sign language and the reallocation of auditory cortex in deaf signers. And I often carried on a sort of mental dialogue with him whenever puzzling problems came up with regard to visual perception or awareness. What, I would wonder, would Francis think of this—how would he attempt to explain it? How would he investigate it?

•

Francis's nonstop creativity—the incandescence that struck me when I first met him in 1986, allied to the way in which he always looked forward, saw years or decades of work ahead for himself and others—made one think of him as immortal. Indeed, well into his eighties, he continued to pour out a stream of brilliant and provocative papers, showing none of the fatigue, or fallings off, or repetitions of old age. It was in some ways a shock, therefore, early in 2003, to learn that he had run into serious medical problems. Perhaps this was in the back of my mind when I wrote to him in May of 2003, but it was not the main reason why I wanted to make contact with him again.

I had found myself thinking of time—time and perception, time and consciousness, time and memory, time and music, time and movement. I had returned, in particular, to the question of whether the apparently continuous passage of time and movement given to us by our eyes was an illusion— whether in fact our visual experience consisted of a series of timeless "moments" which were then welded together by some higher mechanism in the brain. I found myself referring again to the "cinematographic" sequences of stills described to me by migraine patients and which I myself had on occasion experienced. (I had also experienced it very strikingly with other perceptual disorders when I got intoxicated by *sakau* in Micronesia.)

When I mentioned to Ralph that I had started writing about all this, he said, "You have to read Crick and Koch's latest paper. They propose in it that visual awareness really consists of a sequence of 'snapshots'—you are all thinking along the same lines."

I wrote to Francis, enclosing a draft of my article on time. I threw in, for good measure, a copy of my latest book, *Uncle Tungsten*, and some recent articles dealing with our favorite topic of vision. On June 5, 2003, Francis sent me a long letter, full of intellectual fire and cheerfulness and with no hint of his illness. He wrote:

> I have enjoyed reading the account of your early years. I also was helped by an uncle to do some elementary chemistry and glass blowing, though I never had your fascination with metals. Like you I was very impressed by the Periodic Table and by ideas about the structures of the atom. In fact, in my last year at Mill Hill [his school] I gave a talk on how the "Bohr atom," plus quantum mechanics, explained the Periodic Table, though I'm not sure how much of all that I really understood.

I was intrigued by Francis's reactions to *Uncle Tungsten* and wrote back to ask him how much "continuity" he saw between that teenager at Mill Hill who talked about the Bohr atom, the physicist he had become, his later "double helix" self, and his present self. I quoted a letter that Freud had written to Karl Abraham in 1924—Freud was sixty-eight then—in which he had said, "It is making severe demands on the unity of the personality to try and make me identify myself with the author of the paper on the spinal ganglia of the *Petromyzon*. Nevertheless it does seem to be the case."

In Crick's case, the seeming discontinuity was even greater, for Freud was a biologist from the beginning, even though his first interests were in the anatomy of primitive nervous systems. Francis, in contrast, had taken his undergraduate degree in physics, worked on magnetic mines during the war, and went on to do his doctoral work in physical chemistry. Only then, in his thirties—at an age when most researchers are already stuck in what they are doing—did he have a transformation, a "rebirth," as he was later to call it, and turn to biology. In his autobiography, *What Mad Pursuit*, he speaks of the difference between physics and biology:

Natural selection almost always builds on what went before. . . . It is the resulting complexity that makes biological organisms so hard to unscramble. The basic laws of physics can usually be expressed in simple mathematical form, and they are probably the same throughout the universe. The laws of biology, by contrast, are often only broad generalizations, since they describe rather elaborate (chemical) mechanisms that natural selection has evolved over millions of years. . . . I myself knew very little biology, except in a rather general way, till I was over thirty . . . my first degree was in physics. It took me a little time to adjust to the rather different way of thinking necessary in biology. It was almost as if one had to be born again.

By the middle of 2003, Francis's illness was beginning to take its toll, and I began to receive letters from Christof Koch, who by that time was spending several days a week with him. They had become so close, it seemed, that many of their thoughts were dialogic, emerging in the interaction between them, and what Christof wrote to me would condense the thoughts of both. Many of his sentences would start, "Francis and I do have a few more questions about your own experience. . . . Francis thinks this. . . . Myself, I am not sure," and so on.

In response to my paper on time (a version of which was later in *The New York Review of Books* as "In the River of Consciousness"), Crick quizzed me minutely on the rate of visual flicker experienced in migraine auras. These were matters we had discussed when we first met fifteen years earlier, but this, apparently, we had both forgotten; certainly neither of us made any reference to our earlier letters. It was as if no resolution could be reached in 1986, and both of us, in our different ways, had shelved the matter, "forgotten" it, and put it into our unconscious, where it would incubate for another decade and a half before reemerging. Francis and I were converging on a problem which had defeated us before; we were now getting closer to an answer. My feeling of this was so intense in August of 2003 that I felt I had to make a visit to see Francis in La Jolla.

I stayed in La Jolla for a week and made frequent visits to Ralph, who was again working at the Salk. There was a very sweet, noncompetitive air there (or so it seemed to me, as an outsider, in my brief visit), an atmosphere which had delighted Francis when he first arrived at the Salk in the mid-1970s and which had deepened, with his continued presence, ever since. He was still, despite his age, a very central figure there. Ralph

pointed out his car to me, its license plate bearing just four letters, A T G C—the four nucleotides of DNA—and I was happy
to see his tall figure one day going into the lab, still very erect,
though walking slowly, with the aid of a cane.

I made an afternoon presentation one day, and just as I
started, I saw Francis enter and take a seat quietly at the back. I
noticed that his eyes were closed much of the time and thought
he had fallen asleep, but when I finished, he asked a number of
questions so piercing that I realized he had not missed a single
word. His closed-eye appearances had deceived many visitors,
I was told, but they might then find, to their cost, that these
closed eyes veiled the sharpest attention, the clearest and deepest mind, they were ever likely to encounter.

On my last day in La Jolla, when Christof was visiting from
Pasadena, we were all invited to come up to the Crick house
for lunch with Francis and his wife, Odile. "Coming up" was
no idle term; Ralph and I, driving, seemed to ascend continually, around one hairpin bend after another, until we reached
the Crick house. It was a brilliantly sunny California day, and
we all settled down at a table in front of the swimming pool
(a pool where the water was violently blue—not, Francis said,
because of the way the pool was painted, or the sky above it, but
because the local water contained minute particles which, like
dust, diffracted the light). Odile brought us various delicacies—
salmon and shrimp, asparagus—and some special dishes which
Francis, now on chemotherapy, was limited to eating. Though
she did not join the conversation, I knew how closely Odile, an
artist, followed all of Francis's work, if only from the fact that
it was she who had drawn the double helix in the famous 1953
paper and, fifty years later, a frozen runner to illustrate the
snapshot hypothesis in the 2003 paper that had so excited me.

Sitting next to Francis, I could see that his shaggy eyebrows had turned whiter and bushier than ever, and this deepened his sagelike appearance. But this venerable image was constantly belied by his twinkling eyes and mischievous sense of humor. Ralph was eager to tell Francis about his latest work—a new form of optical imaging which could show structures almost down to the cellular level in the living brain. It had never been possible to visualize brain structure and activity on this scale before, and it was on this "meso" scale that both Crick and Gerald Edelman, whatever their differences, now located the functional structures of the brain.

Francis was very excited about Ralph's new technique and his pictures, but at the same time he fired volleys of piercing questions, grilling Ralph, interrogating him, in a minute but also constructive way.

Francis's closest relationship, besides Odile, was clearly with Christof, his "son in science," and it was immensely moving to see how the two men, forty or more years apart in age and so different in temperament and background, had come to respect and love one another so deeply. (Christof is romantically, almost flamboyantly physical, given to dangerous rock climbing and brilliantly colored shirts. Francis seemed almost ascetically cerebral, his thinking so unswayed by emotional biases and considerations that Christof occasionally compared him to Sherlock Holmes.) Francis spoke with great pride, a father's pride, of Christof's forthcoming book, *The Quest for Consciousness*, and then of "all the work we will do after it is published." He outlined the dozens of investigations, years of work, which lay ahead—work especially stemming from the convergence of molecular biology with systems neuroscience. I wondered what Christof was thinking, Ralph too, for it was all too clear to us (and must have been clear to Francis) that his health was declining fast and that he himself would never be able to see more than the beginning of that vast research scheme. Francis, I felt, had no fear of death, but his acceptance of it was tinged with sadness that he would not be alive to see the wonderful, almost unimaginable, scientific achievements of the twenty-first century. The central problem of conscious-ness and its neurobiological basis, he was convinced, would be fully understood, "solved," by 2030. "You will see it," he often said to Ralph, "and you may, Oliver, if you live to my age."

In January of 2004, I received the last letter I would get from Francis. He had read "In the River of Consciousness." "It reads very well," he wrote, "though I think a better title would have been 'Is Consciousness a River?' since the main thrust of the piece is that it may well not be." (I agreed with him.)

"Do come and have lunch again," his letter concluded.

In the mid-1950s, when I was in medical school, there seemed to be an unbridgeable gap between our neurophysiology and the actualities of how patients experienced neurological dis-orders. Neurology continued to follow the clinico-anatomical method set by Broca a century earlier, locating areas of damage in the brain and correlating these with symptoms; thus speech disturbances were correlated with damage to Broca's speech area, paralysis with damage to motor areas, and so on. The brain was regarded as a collection or mosaic of little organs, each with specific functions but somehow interconnected. But there was little idea of how the brain worked as a whole. When I wrote *The Man Who Mistook His Wife for a Hat* in the early

1980s, my thinking was still grounded in this model, in which the nervous system was largely conceived as fixed and invariant, with "pre-dedicated" areas for every function.

Such a model was useful, say, in locating the area of damage in someone with aphasia. But how could it explain learning and the effects of practice? How could it explain the reconstructions and revisions of memory we make throughout our lives? How could it explain the processes of adaptation, of neural plasticity? How could it explain consciousness—its richness, its wholeness, its ever-changing stream, and its many disorders? How could it explain individuality or self?

Although huge advances were being made in the neuroscience of the 1970s and 1980s, there was, in effect, a conceptual crisis or vacuum. There was no general theory which could make sense of the rich data, the observations of a dozen different disciplines from neurology to child development, linguistics, and even psychoanalysis.

•

In 1986, I read a remarkable article by Israel Rosenfield in *The New York Review of Books* in which he discussed the revolutionary work and views of Gerald M. Edelman. Edelman was nothing if not bold. "We are at the beginning of the neuroscientific revolution," he wrote. "At its end, we shall know how the mind works, what governs our nature, and how we know the world."

A few months later, along with Rosenfield, I arranged to meet the man himself, in a conference room near the Rockefeller University where Edelman had his Neurosciences Institute at the time.

Edelman strode in, greeted us briefly, and then talked nonstop for twenty or thirty minutes, outlining his theories; neither of us dared to interrupt him. He then abruptly took his leave, and looking out the window, I could see him walking rapidly down York Avenue, looking to neither side. "That is the walk of a genius, a monomaniac," I thought to myself. "He is like a man possessed." I had a sense of awe and envy—how I should like such a ferocious power of concentration! But then I thought that life might not be entirely easy with such a brain; indeed, Edelman, I was to find, took no holidays, slept little, and was driven, almost bullied, by nonstop thinking; he would often phone Rosenfield in the middle of the night. Perhaps I was better off with my own, more modest endowment.

In 1987, Edelman published *Neural Darwinism*, a seminal volume and the first in a series of books presenting and exploring the ramifications of a very radical idea which he called the theory of neuronal group selection, or, more evocatively, neural Darwinism. I struggled with the book, finding the writing impenetrable at times, in part because of the novelty of Edelman's ideas and in part because of the book's abstractness and lack of concrete examples. Darwin had said of the *Origin* that it was "one long argument," but he buttressed it by innumerable examples of natural (and artificial) selection and by a gift of writing akin to that of a novelist. *Neural Darwinism*, in contrast, was pure argument—a single intense, intellectual brief from beginning to end. I was not the only person to have difficulties with *Neural Darwinism*; the density, audacity, and originality of Edelman's work, its pushing beyond the bounds of language, were daunting.

I annotated my copy of *Neural Darwinism* with clinical examples, wishing that Edelman himself, who had trained as a neurologist and psychiatrist, had done the same.

•

In 1988, I met Gerry again when we both made presentations at a conference on the art of memory in Florence.[3] After the conference, we had dinner together. I found him very different from the monologist I had first encountered, when he had tried to compress a decade of intense thinking into a few minutes; now he was more relaxed, patient with my slowness. And the tone was one of easy conversation. Gerry was eager to learn of my experiences with patients—experiences which might be pertinent to his thinking, clinical stories which might be relevant to his theories of how the brain worked and of consciousness. He was somewhat isolated from clinical life at the Rockefeller, like Crick at the Salk, and both were hungry for clinical data.

Our table had a paper tablecloth, and when points were obscure, we drew diagrams on it until their meaning was fully explored. By the time we had finished, I felt I understood his theory of neuronal group selection, or some of it. It seemed to illuminate a vast field of neurological and psychological knowledge, to be a plausible, testable model of perception, memory, and learning, of how through selective and interactive brain mechanisms a human being achieves consciousness and becomes a unique individual.

•

While Crick (and his co-workers) cracked the genetic code—a set of instructions, in general terms, for building a body— Edelman realized early that the genetic code could not specify

---

3. Gerry had a rapt but puzzled audience, and when he said, "The mind is not a computer, the world is not a piece of tape," his Italian audience misheard this as "The world is not a piece of cake." This led to passionate discussions in the corridors as to what the great American professor meant by this gnomic statement.

or control the fate of every single cell in the body, that cellular development, especially in the nervous system, was subject to all sorts of contingencies—nerve cells could die, could migrate (Edelman spoke of such migrants as "gypsies"), could connect up with each other in unpredictable ways—so that even by the time of birth the fine neural circuitry is quite different even in the brains of identical twins; they are already different individuals who respond to experience in individual ways.

Darwin, studying the morphology of barnacles a century before Crick or Edelman, observed that no two barnacles of the same species were ever exactly the same; biological populations consisted not of identical replicas but of different and distinct individuals. It was upon such a population of variants that natural selection could act, preserving some lineages for posterity, condemning others to extinction (Edelman liked to call natural selection "a huge death machine"). Edelman conceived, almost from the start of his career, that processes analogous to natural selection might be crucial for individual organisms—especially higher animals—in the course of their lives, with life experiences serving to strengthen certain neuronal connections or constellations in the nervous system and to weaken or extinguish others.[4]

Edelman thought of the basic unit of selection and change as being not a single neuron but groups of fifty to a thousand interconnected neurons; thus he called his hypothesis the theory of neuronal group selection. He saw his own work as the completion of Darwin's task, adding selection at a cellular

---

4. Edelman had originally pioneered a selectionist theory in relation to the immune system—he was awarded a Nobel Prize for this work—and then, in the mid-1970s, started to apply analogous concepts to the nervous system.

level within the life span of a single individual to that of natural selection over many generations.

Clearly there are some innate biases or dispositions as part of our genetic programming; otherwise an infant would have no tendencies whatever, would not be moved to do anything, seek anything, to stay alive. These basic biases (towards, for example, food, warmth, and contact with other people) direct a creature's first movements and strivings.

And at an elementary physiological level, there are various sensory and motor givens, from the reflexes that automatically occur (for example, in response to pain) to certain innate mechanisms in the brain (for example, control of respiration and autonomic functions).

But in Edelman's view, very little else is programmed or built in. A baby turtle, on hatching, is ready to go. A human baby is not ready to go; it must create all sorts of perceptual and other categorizations and use them to make sense of the world—to make an individual, personal world of its own, and to find out how to make its way in that world. Experience and experiment are crucially important here—neural Darwinism is essentially *experiential* selection.

The real functional "machinery" of the brain, for Edelman, consists of millions of neuronal groups, organized into larger units or "maps." These maps, continually conversing in ever-changing, unimaginably complex, but always meaningful patterns, may change in minutes or seconds. One is reminded of C. S. Sherrington's poetic evocation of the brain as "an enchanted loom," where "millions of flashing shuttles weave a dissolving pattern, always a meaningful pattern though never an abiding one; a shifting harmony of subpatterns."

The creation of maps that respond selectively to certain elemental categories—for example, to movement or color in the visual world—may involve the synchronization of thousands of neuronal groups. Some mappings take place in discrete and anatomically fixed, pre-dedicated parts of the cerebral cortex, as is the case with color: color is constructed predominantly in the area called V4. But much of the cortex is plastic, pluripotent "real estate" that can serve (within limits) whatever function is needed; thus what would be auditory cortex in hearing people may be reallocated for visual purposes in congenitally deaf people, just as what is normally visual cortex may be used for other sensory functions in the congenitally blind.

Ralph Siegel, in looking at the neural activity in monkeys doing a particular visual task, was very aware of the gulf between "micro" methods, in which electrodes are inserted into a single nerve cell to record its activity, and "macro" methods (fMRI, PET scans, etc.) which show whole areas of the brain responding. Conscious of the need for something in between, he pioneered a very original optical "meso" method that allowed him to look at dozens or hundreds of neurons as they interacted and synchronized with one another in real time. One of his findings—unexpected and baffling at first— was that neuronal constellations or maps could change in a matter of seconds as the animal learned or adapted to different sensory inputs. This was very much in accordance with Edelman's theory of neuronal group selection, and Ralph and I spent many hours discussing the implications of his theory with each other and with Edelman himself, who, like Crick, was fascinated by Ralph's work.

Where perception of objects is concerned, Edelman likes to say, the world is not "labeled"; it does not come "already parsed into objects." We must make our perceptions through our own categorizations. "Every perception is an act of creation," as Edelman says. As we move about, our sense organs take samplings of the world, and from these, maps are created in the brain. There then occurs with experience a selective strengthening of those mappings that correspond to successful perceptions—successful in that they prove the most useful and powerful for the building of "reality."

Edelman speaks here of a further, integrative activity peculiar to more complex nervous systems; this he calls "reentrant signaling." In his terms, the perception of a chair, for example, depends first on the synchronization of activated neuronal groups to form a "map," then a further synchronization of a number of scattered mappings throughout the visual cortex—mappings relating to many different perceptual aspects of the chair (its size, its shape, its color, its "leggedness," its relation to other sorts of chairs—armchairs, rocking chairs, baby chairs, etc.). In this way, a rich and flexible percept of "chairhood" is achieved, which allows the instant recognition of innumerable sorts of chairs *as* chairs. This perceptual generalization is dynamic, so it can be continually updated, and it depends on the active and incessant orchestration of countless details.

Such correlation and synchronization of neuronal firing in widely separated areas of the brain is made possible by very rich connections between the brain's maps—connections which are reciprocal and may contain millions of fibers. Stimuli from, say, touching a chair may affect one set of maps; stimuli from seeing it may affect another set. Reentrant signaling takes place between these sets of maps as part of the process of perceiving a chair.

Categorization is the central task of the brain, and reentrant signaling allows the brain to categorize its own categorizations, then recategorize these, and so on. Such a process is the beginning of an enormous upward path enabling ever higher levels of thought and consciousness.

Reentrant signaling might be likened to a sort of neural United Nations, in which dozens of voices are talking together, while including in their conversations a variety of constantly inflowing reports from the outside world, bringing them together into a larger picture as new information is correlated and new insights emerge.

Edelman, who once planned to be a concert violinist, uses musical metaphors as well. In a BBC radio interview, he said:

> Think: if you had a hundred thousand wires randomly connecting four string quartet players and that, even though they weren't speaking words, signals were going back and forth in all kinds of hidden ways [as you usually get them by the subtle nonverbal interactions between the players] that make the whole set of sounds a unified ensemble. That's how the maps of the brain work by reentry.

The players are connected. Each player, interpreting the music individually, constantly modulates and is modulated by the others. There is no final or "master" interpretation; the music is collectively created, and every performance is unique. This is Edelman's picture of the brain, as an orchestra, an ensemble, but without a conductor, an orchestra which makes its own music.

•

When I walked back to my hotel after dinner with Gerry that evening, I found myself in a sort of rapture. It seemed to me that the moon over the Arno was the most beautiful thing I had ever seen. I had the feeling of having been liberated from decades of epistemological despair—from a world of shallow, irrelevant computer analogies into one full of rich biological meaning, one which corresponded with the reality of brain and mind. Edelman's theory was the first truly global theory of mind and consciousness, the first biological theory of individuality and autonomy.

I thought, "Thank God I have lived to hear this theory." I felt as I imagined many people must have felt in 1859 when the *Origin* came out. The idea of natural selection was astounding but, once one thought about it, obvious. Similarly, when I grasped what Edelman was about that evening, I thought, "How extremely stupid of me not to have thought of this myself!" just as Huxley had said after reading the *Origin*. It all seemed so clear suddenly.

A few weeks after my return from Florence, I had another epiphany, of a rather improbable and comic sort. I was driving up to Lake Jefferson through the lush countryside of Sullivan County, enjoying the tranquil fields and hedgerows, when I saw—a cow! But a cow transfigured by my new Edelmanian view of animal life, a cow whose brain was constantly mapping all its perceptions and movements, a cow whose inner being consisted of categorizations and mappings, neuronal groups flashing and conversing at great speed, an Edelmanian cow suffused by the miracle of primary consciousness. "What a wonderful animal!" I thought to myself. "Never have I seen a cow in this light before."

Natural selection could show me how cows in general had come to be, but neural Darwinism was necessary to apprehend what it was like to be this particular cow. Becoming this particular cow was made possible by experience selecting particular neuronal groups in its brain and amplifying their activity.

·

Mammals, birds, and some reptiles, Edelman speculated, have "primary consciousness," the ability to create mental scenes to help them adapt to complex and changing environments. This achievement, for Edelman, depended upon the emergence of a new type of neuronal circuit at some "transcendent moment" in evolution—a circuit allowing massive, parallel, reciprocal connections between neuronal maps, as well as among the ongoing global mappings that integrate new experiences and recategorize categories.

At some second transcendent moment in evolution, Edelman proposed, the development of "higher-order consciousness" was made possible in humans (and perhaps a few other species including apes and dolphins) by a higher level of reentrant signaling. Higher-order consciousness brings an unprecedented power of generalization and reflection, of recognizing past and future, so that finally self-consciousness, the awareness of being a self in the world, is achieved.

·

In 1992, I went with Gerry to a conference on consciousness at Jesus College in Cambridge. While Gerry's books were often difficult to read, seeing and hearing him speak gave a feeling of revelation to many in the audience.

At that same meeting—I forget what prompted the exchange—Gerry said to me, "You're no theoretician."

"I know," I said, "but I am a field-worker, and you need the sort of fieldwork I do for the sort of theory making you do." Gerry agreed.

•

I often encounter situations in day-to-day neurological practice which completely defeat classical neurological explanations and cry out for explanations of a radically different kind, but many such phenomena can be explained, in Edelmanian terms, as breakdowns in local or higher-order mapping consequent upon nerve damage or disease.

When, after injury and immobility following my accident in Norway, my left leg became "alien," my neurological knowledge did not help; classical neurology had nothing to say about the relation of sensation to knowledge and to self, about how, if the flow of neural information is impaired, a limb may be lost to consciousness and to self, "disowned," and how there may then be a rapid remapping of the rest of the body which excludes that limb.

If the right hemisphere of the brain is badly damaged in its sensory (parietal) areas, patients may show an "anosognosia," an unawareness that anything is the matter, even though the left side of their body is senseless or paralyzed. Sometimes they may insist that their own left side belongs to "someone else." For such patients, subjectively, their space and world are entire, even though they are living in a hemi-world. For many years, anosognosia was misinterpreted as a bizarre neurotic symptom, since it was unintelligible in terms of classical neurology. But Edelman sees such a condition as a "disease of consciousness," a total breakdown of high-level reentrant signaling and mapping in one hemisphere and a radical reorganization of consciousness in consequence.

Sometimes following a neurological lesion, a dissociation occurs between memory and consciousness, leaving only implicit knowledge or memory. Thus my patient Jimmie, the amnesic mariner, had no explicit memory of Kennedy's assassination, and when I asked him whether any president had been assassinated in the twentieth century, he would say, "No, not that I know of." But if I asked him, "Hypothetically, then, if a presidential assassination had somehow occurred without your knowledge, where might you guess it occurred: New York, Chicago, Dallas, New Orleans, or San Francisco?" he would invariably "guess" correctly, Dallas.

Similarly, patients with total cortical blindness due to massive damage to the primary visual areas of the brain will assert that they can see nothing, but they may also mysteriously "guess" correctly what lies before them—so-called blindsight. In all these cases, perception and perceptual categorization have been preserved but have been divorced from higher-order consciousness.

Individuality is deeply imbued in us from the very start, at the neuronal level. Even at a motor level, researchers have shown, an infant does not follow a set pattern of learning to walk or how to reach for something. Each baby experiments with different ways of reaching for objects and over the course of several months discovers or selects his own motor solutions. When we try to envisage the neural basis of such individual learning, we might imagine a "population" of movements (and their neural correlates) being strengthened or pruned away by experience.

Similar considerations arise with regard to recovery and rehabilitation after strokes and other injuries. There are no rules; there is no prescribed path of recovery; every patient must discover or create his own motor and perceptual patterns, his own solutions to the challenges that face him; and it is the function of a sensitive therapist to help him in this.

And in its broadest sense, neural Darwinism implies that we are destined, whether we wish it or not, to a life of particularity and self-development, to make our own individual paths through life.

•

When I read *Neural Darwinism*, I wondered if it would change the face of neuroscience as Darwin's theory had changed the face of biology. The short, but inadequate, answer is that it has not, although there are now countless scientists who take for granted many of Edelman's ideas without acknowledging, or perhaps even knowing, that they *are* Edelman's. In this sense, his thinking, though not explicitly acknowledged, has been shifting the very grounds of neuroscience.

In the 1980s, Edelman's theory was so novel that it could not easily be fitted into any of the existing models, the paradigms, of neuroscience, and it was this, I think, which prevented its wide acceptance—coupled with Edelman's at times dense and difficult writing. Edelman's theory was "premature," so ahead of its time, so complex and demanding of new ways of thinking, that it was resisted, or ignored, in the 1980s, but in the next twenty or thirty years, with new technologies, we will be in a good position to verify (or disprove) its ultimate tenets. It remains, for me, the most powerful and elegant explanation of how we humans and our brains construct our very individual selves and worlds.

Home

I sometimes felt that I had left England in an underhand way. I had had the best of English educations, had absorbed the best of English diction and prose, the habits and traditions of a thousand years, and here I was taking this precious mental cargo, everything that had been invested in me, out of the country, without so much as a "Thank you" or a "Good-bye."

Nonetheless, I continued to regard England as home, returned as frequently as I could, and felt stronger—a better writer—whenever my feet were planted on home ground. I kept in close touch with my relatives, friends, and colleagues in England and made belief that my ten, my twenty, my thirty years in America were nothing more than an extended visit and that, sooner or later, I would return home.

My sense of England as "home" took a beating in 1990, when my father died and the house on Mapesbury Road—where I was born and brought up and which I revisited and often stayed in when I returned to England, the house of which every inch was suffused for me with memories and emotions—was sold. I no longer felt I had a place to go back to, and my visits thereafter felt like visits and not like returns to my own country and people.

I was strangely proud, nonetheless, of my U.K. passport, which (prior to 2000) had beautiful large, stiff boards embossed with gold lettering, very different from the flimsy little things issued by most countries. I did not seek American citizenship and was happy to have a green card, to be accounted a "resident alien." This accorded with how I felt, at least for much of the time—a friendly, observant alien noting everything around me but without civic responsibilities such as voting or jury duty or need to affiliate myself with the country's policies or politics. I often felt (as Temple Grandin said of herself) that I was an anthropologist on Mars. (I had much less of this feeling in my California days, when I felt at one with the mountains, the forests, and the deserts of the West.)

And then, in June of 2008, to my surprise, I heard that my name was in the Queen's Birthday Honours List—that I was to become a Commander of the Order of the British Empire. The term "commander" tickled me—I could not imagine myself as a commander on the bridge of a destroyer or a battleship—but I was curiously and rather deeply moved by the honor.

Though I am not given to formal clothes or other formalities—normally, my clothes are sloppy and decrepit, and I have only one suit—I enjoyed the formalities in Buckingham Palace: being instructed how to bow, how to walk backwards before the queen, how to await her taking one's hand or addressing one. (The royal person was not to be touched, or spoken to, unbidden.) I was half-afraid that I would do something awful, like faint or fart right in front of the queen, but all went well. During the ceremony, I was very impressed by the queen's stamina: by the time I was called up, she had been standing erect, without support, for more than two hours (there were two hundred honorees that day). She spoke to me briefly but warmly, asking me what I was working on. I had the feeling of a very decent, friendly person with a sense of humor. It was as if she—and England—were saying, "You have done useful, honorable work. Come home. All is forgiven."

Medical life, seeing patients, was not displaced by the writing of *Seeing Voices*, of *Island*, or of *Uncle Tungsten*. I continued to see patients at Beth Abraham, the Little Sisters, and elsewhere.

In the summer of 2005, I went to England to visit Clive Wearing, the extraordinary amnesic musician who had been the subject of Jonathan Miller's 1986 film, *Prisoner of Consciousness*. Clive's wife, Deborah (with whom I had corresponded over the years), had just published a remarkable book about him, and she hoped I might now see what he was like twenty years after his disastrous encephalitis. While he could remember almost nothing of his adult life and could hold on to new events for no more than a few seconds, Clive could still play the organ and conduct a choir, just as he had once done as a professional musician. He illustrated the special power of music and of musical memory, and I wanted to write about this. Thinking about this and many other "neuromusical" topics, I felt I should try to put together a book on music and the brain.

Musicophilia, as the book would eventually be called, started out as a modest project; I thought it might be a tiny book, perhaps three chapters. But as I started thinking about people with musical synesthesia; people with amusia, who could not recognize any music; people with frontotemporal dementia who might have a sudden burst or release of unsuspected musical talents and passions; people with musical seizures, or seizures induced by music; and people haunted by "earworms" or repetitive images of music or outright hallucinations of music, the book grew much larger.

I had been fascinated, moreover, by the therapeutic power of music since seeing it in my postencephalitic patients forty years before, even before they were awakened by L-dopa. Since then, I had been struck by the powers of music to help patients with many other conditions: amnesia, aphasia, depression, and even dementia.

Since *Hat* was first published in 1985, I have received an ever-increasing flow of letters from readers, often offering descriptions of their own experiences. This has extended my practice, so to speak, far beyond the confines of the clinic. *Musicophilia* (and later *Hallucinations*) was greatly enriched by some of these letters and reports, no less than by my correspondence and visits with physicians and researchers.

And while I wrote about many new patients and subjects in *Musicophilia*, I also revisited a number of the patients I had written about earlier, this time focusing on their responses to music and considering them in the light of new forms of brain imaging and concepts of how the brain-mind creates constructs and categories.

•

As I entered my seventies, I was in excellent health; I had a few orthopedic problems but nothing serious or life threatening. I did not give much thought to illness or death, even though I had lost all three of my older brothers, as well as many friends and contemporaries.

In December of 2005, however, a cancer made its presence suddenly and dramatically known—a melanoma in my right eye, which presented as a sudden incandescence to one side and then a partial blindness. It had probably been growing slowly for some time and at this point had reached close to the fovea, the tiny central area where vision is most acute. Melanoma has an evil reputation, and when the diagnosis was pronounced, I took it as a death sentence. But ocular melanomas, my doctor quickly said, were relatively benign; they rarely metastasized, and they were eminently treatable.

The cancer was irradiated, then lasered several times, because certain areas kept regrowing. During the first eighteen months of treatment, sight fluctuated in my right eye almost daily, from near blind to near normal, and I would be thrown, with these fluctuations, from terror to relief, then back to terror—from one emotional extreme to another.

This would have been hard to bear (and I would have been even harder to live with) had I not become fascinated by some of the visual phenomena which occurred as, bit by bit, my retina—and eyesight—were nibbled away by the tumor and the lasering: the wild topological distortions, the perversions of color, the clever but automatic filling in of blind spots, the incontinent spread of color and form, the continued perception of objects and scenes when the eyes were closed, and, not least, the varied hallucinations which now swarmed in my ever-larger blind spots. My brain was clearly as involved as the eye itself.

I feared blindness, but I feared death even more, so I made a sort of bargain with the melanoma: take the eye if you must, I told it, but leave the rest of me alone.

In September of 2009, after three and a half years of treatment, the retina in my right eye, fragile from radiation, hemorrhaged, blinding it completely; attempts to remove the blood failed, because the retina immediately re-bled. Without binocular vision, I now had many new, disabling (but sometimes enthralling!) phenomena to contend with—and investigate. The loss of stereo vision was not only a sad privation for me as a passionate stereophile but often dangerous. Without depth perception, steps and curbs just looked like lines on the ground, and distant objects seemed to be on the same plane as closer ones. With the loss of visual field to my right, I had many accidents, colliding with objects or people who seemed to loom suddenly in front of me out of nowhere. And I was not only physically blind but mentally blind, to the right. I could no longer even *imagine* the presence of what I could no longer see. Such a hemi-neglect, as neurologists call it, is usually the result of a stroke or a tumor in the visual or parietal areas of the brain. For me, as a neurologist, these phenomena were especially fascinating, for they provided an astonishing panorama of the ways in which the brain works (or misworks, or fails to work) when the input from the senses is deficient or abnormal. I documented all this in minute detail—my melanoma journals ran to ninety thousand words—and studied it, performing perceptual experiments of all sorts. The whole experience, like my "leg" experience, became an *experimentum suitatis*, an experiment with, or on, myself.

The perceptual consequences of my eye damage constituted a fertile ground of enquiry; I felt as if I were discovering a whole world of strange phenomena, although I could not help thinking that all patients with eye problems like mine surely experienced some of the same perceptual phenomena as I did. Writing of my own experiences, then, I would also be writing for them. But the sense of discovery was exhilarating and kept me going through what might otherwise have been rather fearful and demoralizing years, as did my continuing seeing patients and writing.

·

I was working hard on a new book, *The Mind's Eye*, when a new series of mishaps and surgical challenges hit me. In September of 2009, immediately after the hemorrhage in my right eye, I had to have a total replacement of my left knee (this too, of course, generated a modest journal). I was told that I had a window of eight weeks or so after the surgery to regain full range of motion at the knee; if I did not succeed in this, I would have a stiff leg for the rest of my life. Working the knee, tearing down scar tissue, would be very painful. "Don't be brave," the surgeon said. "You can have all the painkillers you need." My therapists, moreover, spoke of the pain in almost amorous terms. "Embrace it," they said. "Sink into it." It was "good pain," they insisted, and pushing myself to the limits was crucial if I was to gain full flexibility in the short window I had.

I was doing nicely in rehab, gaining range of motion and strength by the day, when an unwelcome further problem hit me: the sciatica I had battled for many years reemerged, slowly, slyly, at first, but rapidly reaching an intensity beyond anything I had known before.

I struggled to continue the rehab, to keep active, but the sciatic pain beat me down, and by December I was bedridden. I had lots of morphine left from my knee surgery—it had been invaluable helping with the "good" knee pain—but it was virtually useless against the neuralgic pain typical of a crushed spinal nerve. (This is so of all "neuropathic" pain.) It became absolutely impossible to sit, even for a second.

I was unable to sit and play the piano—a severe deprivation, because I had returned to piano-playing and music lessons when I had turned seventy-five (having written about how even older people can learn new skills, I thought it was time to take my own advice). I tried to play standing up, but I found this impossible.

I did all my writing standing up; I made a special high platform on my worktable, using ten volumes of the *Oxford English*

Dictionary as props. The concentration involved in writing, I found, was almost as good as the morphine and had no side effects. I hated lying in bed, in a hell of pain, and spent as many hours as I could writing at my improvised standing desk.

Some of my thinking and writing and reading at this time, indeed, was *about* pain, a subject I had never really thought about. My own recent experience, in the course of two months, had shown me that there were at least two radically different sorts of pain. The pain from my knee surgery was entirely local; it did not spread beyond the knee area and was entirely dependent on how much I stretched the operated-upon and contracted tissues. I could easily quantify it on a ten-point scale, and, above all, as the therapists said, it was "good pain," pain one could embrace, work through, and conquer.

The "sciatica" (an inadequate term) was wholly different in quality. It was not local, as a start; it spread far beyond the area innervated by my impinged L5 nerve roots on the right. It was not a predictable response to the stimulus of stretching, as the knee pain was. Instead, it came in sudden paroxysms that were quite unpredictable and could not be prepared for; one could not grit one's teeth in advance. Its intensity was off the scale; there was no quantifying it; it was, simply, overwhelming.

Even worse, this sort of pain had an affective component all its own, which I found difficult to describe, a quality of agony, of anguish, of horror—words which still do not catch its essence. Neuralgic pain cannot be "embraced," fought against, or accommodated. It crushes one into a quivering, almost mindless sort of pulp; all one's powers of will, one's very identity, disappear under the assault of such pain.

I reread Henry Head's great *Studies in Neurology*, where he contrasts "epicritic" sensations—precisely localized, discrimi-natory, and proportional to stimulation—with "protopathic" sensations: diffuse, affect laden, paroxysmal. This dichotomy seemed to correspond well to the two types of pain I had expe-rienced, and I wondered about writing a little, very personal book or essay on pain, resurrecting, among other things, Head's long-forgotten terms and distinctions. (I inflicted my thoughts at length on friends and colleagues, but I never completed the essay I intended.)

By December, my sciatica had become so overwhelming that I could no longer read or think or write and, for the first time in my life, thought about suicide.[1]

Spinal surgery was scheduled for December 8. I was on huge doses of morphine at this point, and my surgeon had warned me that the pain might become even worse, from postoperative edema, for a couple of weeks after surgery—as, indeed, turned out to be the case. December of 2009, then, continued very grim, and perhaps the heavy medications I was taking for pain heightened all the feelings I had at the time, the often sudden shifts between hope and fear.

Unable to bear twenty-four hours a day on my bed but still needing to lie down, I started (a stick in one hand, holding Kate's arm with the other) to make my way to the office, where I could at least dictate letters and answer phone calls, make believe I was back at work, while lying on the office couch.

---

1. My friend and colleague Peter Jannetta—we were residents together at UCLA—was able to make a discovery, and perfect a technique, which completely altered and often saved the lives of people with trigeminal neuralgia, a paroxysmal pain in the eye and face which (before Peter's work) had no remedy, was often "beyond bearing," and not infrequently led to suicide.

Shortly after my seventy-fifth birthday in 2008, I met someone I liked. Billy, a writer, had just moved from San Francisco to New York, and we began having dinners together. Timid and inhibited all my life, I let a friendship and intimacy grow between us, perhaps without fully realizing its depth. Only in December of 2009, still recuperating from knee and back surgeries and racked with pain, did I realize how deep it was.

Billy was going to Seattle to spend Christmas with his family, and just before he went, he came to see me and (in the serious, careful way he has) said, "I have conceived a deep love for you." I realized, when he said this, what I had not realized, or had concealed from myself before—that I had conceived a deep love for him too—and my eyes filled with tears. He kissed me, and then he was gone.

I thought of him almost constantly while he was away, but not wanting to disturb him while he was with his family, I awaited his phone calls with intense eagerness combined with a sort of tremulousness. On the days when he could not phone me at the usual time, I became terrified that he had been disabled or killed in a traffic accident and almost sobbed with relief when, an hour or two later, he called.

There was an intense emotionality at this time: music I loved, or the long golden sunlight of late afternoon, would set me weeping. I was not sure what I was weeping for, but I would feel an intense sense of love, death, and transience, inseparably mixed.

Lying in bed, I kept a notebook of all my feelings—a notebook devoted to "falling in love." Billy came back late on the evening of December 31, bringing a bottle of champagne. We toasted each other when he opened the bottle, each saying, "To you." And then we toasted the New Year as it came in.

•

In the last week of December, the nerve pain had started to grow less. Was this because the postoperative edema was settling? Or was it—a hypothesis I could not help entertaining— because the joy of being in love was a match for the pain of the neuralgia and could alleviate it almost as well as Dilaudid or fentanyl? Did being in love itself flood the body with opioids, or cannabinoids, or whatever?

By January, I was back to writing on my improvised desk of *OED*s, and I was now able to go out a bit, provided I could stand. I stood at the back of concert halls and lecture halls, went to restaurants if they had a bar I could stand at, and resumed seeing my analyst, though I had to stand when I faced him in his consulting room. I went back to the manuscript of *The Mind's Eye*, abandoned on my desk when I became bed-bound.

•

It has sometimes seemed to me that I have lived at a certain distance from life. This changed when Billy and I fell in love. As a twenty-year-old, I had fallen in love with Richard Selig; as a twenty-seven-year-old, tantalizingly, with Mel; as a thirty-two-year-old, ambiguously, with Karl; and now (for God's sake!) I was in my seventy-seventh year.

Deep, almost geological changes had to occur; in my case, the habits of a lifetime's solitude, and a sort of implicit selfishness and self-absorption, had to change. New needs, new fears, enter one's life—the need for another, the fear of abandonment. There have to be deep, mutual adaptations.

For Billy and me, these were made easier by shared interests and activities; we are both writers, and this, indeed, is how we met. I had read proofs of Billy's book *The Anatomist* and admired it. I wrote to him and suggested that we might meet if he found himself on the East Coast (which he did, on a visit to New York in September of 2008). I liked his thinking, which was both serious and playful, his sensitivity to the feelings of others, and his combination of forthrightness and delicacy. It was a new experience for me to lie quietly in someone's arms and talk, or listen to music, or be silent, together. We learned to cook and eat proper meals together; I had more or less lived on cereal up to this point, or sardines, which I would eat out of the tin, standing up, in thirty seconds. We started to go out together—sometimes to concerts (which I favored), sometimes to art galleries (which he favored), and often to the New York Botanical Garden, which I had traipsed around, alone, for more than forty years. And we started to travel together: to my city, London, where I introduced him to friends and family; to his city, San Francisco, where he has many friends; and to Iceland, for which we both have a passion.

We often swim together, at home or abroad. We sometimes read our works in progress to each other, but mostly, like any other couple, we talk about what we are reading, we watch old movies on television, we watch the sunset together or share sandwiches for lunch. We have a tranquil, many-dimensional sharing of lives—a great and unexpected gift in my old age, after a lifetime of keeping at a distance.

They called me Inky as a boy, and I still seem to get as ink stained as I did seventy years ago.

I started keeping journals when I was fourteen and at last count had nearly a thousand. They come in all shapes and sizes, from little pocket ones which I carry around with me to enormous tomes. I always keep a notebook by my bedside, for dreams as well as nighttime thoughts, and I try to have one by the swimming pool or the lakeside or the seashore; swimming too is very productive of thoughts which I must write, especially if they present themselves, as they sometimes do, in the form of whole sentences or paragraphs.

When writing my *Leg* book, I drew heavily on the detailed journals I had kept as a patient in 1974. *Oaxaca Journal*, too, relied heavily on my handwritten notebooks. But for the most part, I rarely look at the journals I have kept for the greater part of a lifetime. The act of writing is itself enough; it serves to clarify my thoughts and feelings. The act of writing is an integral part of my mental life; ideas emerge, are shaped, in the act of writing.

My journals are not written for others, nor do I usually look at them myself, but they are a special, indispensable form of talking to myself.

The need to think on paper is not confined to notebooks. It spreads onto the backs of envelopes, menus, whatever scraps of paper are at hand. And I often transcribe quotations I like, writing or typing them on pieces of brightly colored paper and pinning them to a bulletin board. When I lived in City Island, my office was full of quotations, bound together with binder rings that I would hang to the curtain rods above my desk.

Correspondence is also a major part of life. On the whole, I enjoy writing and receiving letters—it is an intercourse with other people, *particular* others—and I often find myself able to write letters when I cannot "write," whatever Writing (with a capital *W*) means. I keep all the letters I receive, as well as copies of my own. Now, trying to reconstruct parts of my life—

such as the very crucial, eventful time when I came to America in 1960—I find these old letters a great treasure, a corrective to the deceits of memory and fantasy.

A vast amount of writing has gone into my clinical notes—and for many years. With a population of five hundred patients at Beth Abraham, three hundred residents in the Little Sisters homes, and thousands of patients in and out of Bronx State Hospital, I wrote well over a thousand notes a year for many decades, and I enjoyed this; my notes were lengthy and detailed, and they sometimes read, others said, like novels.

I am a storyteller, for better and for worse. I suspect that a feeling for stories, for narrative, is a universal human disposition, going with our powers of language, consciousness of self, and autobiographical memory.

The act of writing, when it goes well, gives me a pleasure, a joy, unlike any other. It takes me to another place—irrespective of my subject—where I am totally absorbed and oblivious to distracting thoughts, worries, preoccupations, or indeed the passage of time. In those rare, heavenly states of mind, I may write nonstop until I can no longer see the paper. Only then do I realize that evening has come and that I have been writing all day.

Over a lifetime, I have written millions of words, but the act of writing seems as fresh, and as much fun, as when I started it nearly seventy years ago.

Acknowledgments

Putting this autobiography together would not have been possible without Kate Edgar. Kate has played a unique role in my life—as personal assistant, editor, collaborator, and friend—for more than thirty years (I dedicated my last book, *Hallucinations*, to her). And here, with the help of our two devoted assistants, Hallie Parker and Hailey Wojcik, she has helped me sift through all my earlier writings, published and unpublished, as well as notebooks and letters going back to the 1950s.

I owe a special debt to my friend and fellow neurologist Orrin Devinsky, with whom I have enjoyed a doctor-to-doctor as well as a friend-to-friend dialogue for twenty-five years. Orrin has cast a critical eye on the scientific and clinical parts of this book, as of several earlier books (he was a dedicatee of *Musicophilia*).

Dan Frank, my editor at Knopf, has read through successive versions of this book, offering precious advice and insights at each stage.

My dear friend (and fellow writer) Billy Hayes has been intimately concerned with the genesis, the writing, and the shaping of this book, and it is to him that I dedicate it.

Hundreds of people have been dear and important to me in the course of a long and eventful life, but only a few of them could be brought into the compass of this book. The others should be assured that I have not forgotten them, and that they will reside in my memory and my affections until the day I die.

Index

MUSICOPHILIA
Tales of Music and the Brain

In *Musicophilia*, Oliver Sacks shows us a variety of what he calls "musical misalignments." Among them: a man struck by lightning who suddenly desires to become a pianist at the age of forty-two; an entire group of children with Williams syndrome, who are hypermusical from birth; people with "amusia," to whom a symphony sounds like the clattering of pots and pans; and a man whose memory spans only seven seconds—for everything but music. Illuminating, inspiring, and utterly unforgettable, *Musicophilia* is a masterpiece.

Psychology

ALSO AVAILABLE

An Anthropologist on Mars
The Island of the Colorblind
Migraine
The Mind's Eye
Oaxaca Journal
Seeing Voices
Uncle Tungsten

VINTAGE BOOKS
Available wherever books are sold.
www.vintagebooks.com